THE MISSION IS SALES

THE GOAL IS MORE SALES

Digital Sales Strategies for
"Small & Medium-Sized Businesses"

Authors: Robert Dumouchel & Priyanka Singh

www.smsrd.com

2021

Library of Congress Cataloging-in-Publication Data

Dumouchel, Robert, author.
2021 The Mission is Sales: The Goal is more Sales

ISBN 978-0-9982139-4-1
1. Digital Sales Title.   V7.04

Library of Congress Control Number: 2021907411

WELCOME!
To your journey in Digital Marketing

If you want to improve the digital sales performance of a small and medium-sized business (SMB), then we want to be your marketing Yoda. For many SMB's marketing is a riddle wrapped in an enigma conjured up by black magic. Our goal is to give you the tools to unravel marketing and teach you how to use marketing to create an unfair advantage for your business.

Sales and Marketing conceptually share concepts with the military. They both run campaigns with strategies and tactics, and they both must-win battles and the hearts and minds of the population. The military's campaigns and tactics happen in the fog of war. Sales and Marketing campaigns and tactics happen in the fog of the market. Our brand, "The Art of Word War," is advice you can trust on your journey to market domination.

This is not a step-by-step how-to book because digital marketing changes at a mind-numbing pace. Our goal is to communicate the principles to improve your mastery of marketing. There are always people that can help you with the details, but only you should control the strategy of your journey.

Throughout this book, we tell stories because people remember stories. Stories help us communicate the moral of the story. We jealously guard our client's privacy, so we change businesses, persons, and industry names in our stories but not the facts and circumstances.

All businesses, at the most basic level, share the same mission and goal. **The Mission is Sales. The Goal is more Sales.** Everything else comes after that. We use industry examples, but the concepts apply to any SMB (Small or Medium-Sized Business).

Table of Contents

Word Wars®

If your website address starts with www then, your business is locked in a Word War with your competitors. Like all wars, the winner will be the one that wins, not just the battles but also the hearts and minds of the people. As Iraq taught us, it is amazingly easy to win the battle and lose the war. The same is true in business.

Your weapons are your words and images and how you creatively and effectively deploy them. In the context of a Word War, words include anything that communicates meaning to the market. This includes text, graphics, video, and audio. If it conveys a meaning to the market, then conceptually, it is a word.

Word Wars are waged in search engines, social media, email, and any other form that communicates with the market. We share the same language, so every business has the same weapons. What creates the winner is their creative use of strategy and tactics. Included in this is how you weave together the campaigns, battles, and follow-through.

Word Wars utilized earned and paid weapons. Data analysis is how we evaluate their effectiveness. Advertisements are **paid** traffic, while organic (SEO) traffic is **earned**. Beginners often confuse earned with free. SEO is only free if you have a source of free professional labor. Advertisements are paid for, but they consume more than just the placement cost. There is creative labor on both sides that must be accounted for. Earned traffic varies from Paid traffic in the ongoing placement cost, but both have a one-time creative cost.

All business challenges can be solved with creativity or cash; the choice is yours. Small businesses normally have a shortage of cash and an abundance of creativity. Spend your creativity first and save your cash for those things you cannot earn. Creativity is not free because it requires labor, and your time should never be considered free.

Word War Basic Training

Welcome to Word Warrior basic training. This training is an exhausting and exhilarating experience that will push you to your physical and physiological limits. The goal of the book is to do the same thing to your digital sales and marketing skills.

Armies have basic training to supply a universally shared experience and primary skills everyone can count on. Therefore you need basic training in your Sales and Marketing Teams. This training instantly gives you something in common with everyone you serve with. During this time, your Drill Sergeant will give you everything you need and expect everything in return from you. In the Word War Army, you will learn the same thing - all the basics.

Armies use a repetitive training method until it becomes part of you. This training converts the task from something you think about to something that you just instinctively do. This chapter is muscle-memory training, and the items in this section need to reach that level.

Read this until you MASTER it.

Word Wars are fought first and foremost in Search Engines, and you must learn all about this battleground to survive combat and take down your enemy. The battleground divides into sections, and each has its priorities. Winning in one does not make you a winner in the others. In a broad sense, they divide into the concept of Paid and Earned positions, and almost universally, being first is the goal.

SERP Page Sections:

Section Name	Category
Local Services Ads	Paid
Shopping	Hybrid - Paid & Earned
Google Ads aka AdWords	Paid
Maps	Hybrid – Paid & Earned
Organic	Earned
Images	Earned
Video	Earned

Local Service Ads are separate from Google Ads and have a full implementation and verification. Currently, this is a test, but it is widely available in many industries. The pricing on this is currently a fixed price, but there is a test under way to move this to auction pricing.

Shopping Ads are paid positions. The ads include item pictures and prices connected through the Merchant Center. Optimization is based on the product description from your data feed.

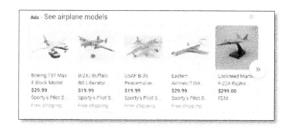

Google Ads are the standard paid text ads placed by the Google Ads System. Optimization is based on bid and quality scores.

Map listing often called the three Pack, is a hybrid section that includes earned and paid. When they both exist, the ads will be in the top positions. Optimization is based on reviews and keyword optimization.

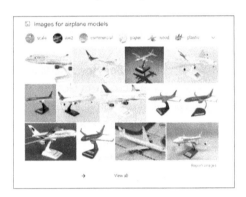

Images are based on the keyword or from image analysis. If the photo is used with an Alt Tag description, that will affect your score on the photo. Optimization rules are vague, but there is agreement that the meta tags and alt tags are important parts.

Organic is where everything started, and it is where earned competition begins. Optimization is based on relevancy to the query, page rank, site reputation, freshness, keyword density, and more than a little magic. Fully 25% of this book is about optimization in this section. For your basic training, make sure you understand these big concepts.

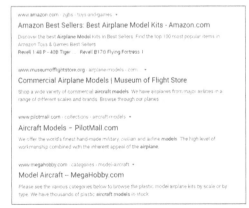

The Video section is largely based on the Meta Description of the video, placement in networks, and other measurements. These are a hybrid section with both paid and earned positions on the page.

The SERP (Search Engine Results Page) is a collection of sections based on Google's decision of what will best serve the searcher. As a trainee, you must identify each section and describe the optimization concepts used for these listings. The sections may appear in any order, but they will appear in this order most of the time. There is no guarantee that any section will appear in a specific search.

Basic Formulas

You do not have to be a math wizard for basic training, but you need to understand the basics. Formulas will be covered in-depth later in the book, but as a Word Warrior in Basic Training, you must master the concepts. We assure you that the math can be done with a calculator and a few addresses for basic tools. What you do with the results is what is important.

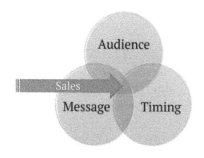

Probability of a Sale

$$P(y) = a \cap m \cap t$$

Sales happen at the **intersection** $\cap$ of the right **a**udience, the right **m**essage, and the right **t**iming. Your job as a Word Warrior is to manage these data sets and bring the magic together.

The stronger those intersections are, the higher the probability of a sale. The key to driving sales is to increase the strength of these factors. Your strategies and tactics should always understand these relationships.

Page Rank

$$PR(A) = (1-d) + d\ (PR(T1)/C(T1) \ldots\ PR(Tn)/C(Tn))$$

Okay, this one does look scary, but it is not. The page rank formula is what makes Google Google. The page rank is the sum of the page rank of the pages that points to it. This calculation runs multiple times (iterates), and on each cycle, it gets more accurate but never reaches perfection.

Weighted Keyword Density

$$density = \Sigma\ (occurrences * position\ weight)$$

Keyword density is the value of that word for the page it occurs on and position counts. It makes an enormous difference if the word is in the page title compared to the body text. Position weights apply to Page Titles, Meta Descriptions, H1 to H6 tags, body text, links, and lists. Everyone, including your favorite author, is guessing about the actual weight, but there is no debate about the concept.

Google Magic

$$Page\ Score = (Word * W1) * (Page * W2) * (CTR * W3) * Overrides$$

The magic of Google brings the various attribute classes of the Word Density, Page Rank, and CTR. The W1-3 are the weights of each section, and in total, they equal 1. You appear on the Organic section on the SERP based on the Word Density, Page Rank, CTR, and any manual overrides. A high density of the keyword on a highly ranked page, with a good CTR history and no manual overrides, will win the coveted first position. To make this easier to understand, we think of the page rank and CTR as the points earned and the word density as the distribution plan for those points.

VOA = Volume of Actions

The Volume of Actions (VOA) is a fundamental measurement of actions. An action is any event the business has **"deemed of value."** It is common for businesses to have more than one Action that they track. Every business values new customers, but they also value email subscriptions, sales inquiries, phone calls, or engaged readers of certain content. In the definition of action, the business should set the value.

CPA = Cost Per Action
CPA = Period Cost/Period VOA

A simple formula that anyone with a calculator can perform with no real effort. The CPA is a basic marketing measurement, and it comes in different time slices and qualifiers. CPAs are qualified as a target, plan, actual, or maximum. Then commonly compared to each other. Examples of this would be the comparison of the Target-CPA to the Actual CPA. It is also common to time slices the CPA data. Examples of this are hourly, daily, weekly, monthly, or annually. Target is your goal, while Plan is what you need to meet your business plan goals. Maximum is the point where you have to stop what you are doing and reassess your plan.

Paid or Earned but Never Free

Conceptually the Internet is the world's largest and most complex collection of publishing businesses. As a Publisher, each website has an editorial policy that balances editorial content with advertising space. There are two incredibly complex simple issues this policy must address the are:

1. Too high and you do not attract visitors
2. Too low and you go bankrupt

Google is a splendid example of this. Their search engine results are why people go to Google, but the ads pay for the 114,096 people (as of Q3 2019) it takes to run Google. If you think that Google is a free service, you do not understand the business model. All space on the internet is either earned or paid.

You earn a position when you create content that people find of value. If you create that value, then it is possible to monetize that by placing advertising next to that. This ad placement is called Display Advertising.

Search or Display Advertising

Advertising comes in two flavors, search or display. Just like a flavor, they are not better or worse. They are simply different.

Search Advertising is driven by the search query submitted by the searcher to the search engine. The search query is what you asked for. The searcher is you, and the search engine is where you submitted the query. You get back the SERP, Search Engine Results Page, a mix of editorial and advertising space.

Display Advertising is displayed next to content that attracted the reader to the website. Display ads are targeted by keywords, audiences, or direct placements. The concept is that the content qualifies the audience. Display advertising is something users run into rather than requesting, as they do in search. Because of the lack of user intent, Display Advertising is typically less responsive and less expensive.

Moving People to Action

Calling people to action is the grand goal of this book and the grand challenge of your business. Sales and Marketing is not a simple action-reaction relationship. But, each touch in the path needs to use one of the weapons of influence listed below. For your basic training, so you need to memorize the Weapons of Influence and understand their application.

People act when you use one of these Weapons of Influence:

1. Reciprocity
2. Commitment
3. Consistency
4. Social proof
5. Authority
6. Liking
7. Scarcity
8. Deals
9. Trust
10. Influencers

> # Commit this list
> # to memory.

These attributes are discussed in detail in the "**Weapons of Influence**" chapter of this book.

Understanding Google

Google surprises people sometimes because they do not understand Google's value system. Many people think that money is the only thing that motivates a business. In Google's case, this is not true. They value the quality of the SERP much more than the money you offer. We have seen time after time that Google is faithful to this value. They know that the billions they make each year are driven by the quality of the SERP because that is what brings billions of people back to Google for their next search. Google understands that the return visit drives their value.

After the Click

This book focuses on getting the click, but that is not the end of the story. After the click comes to the next critical stage known as the "Customer Experience." It is only with a great customer experience that you will get value from the click you just bought or earned. Make sure that it is

an experience worth talking about

The Sales Ecosystem

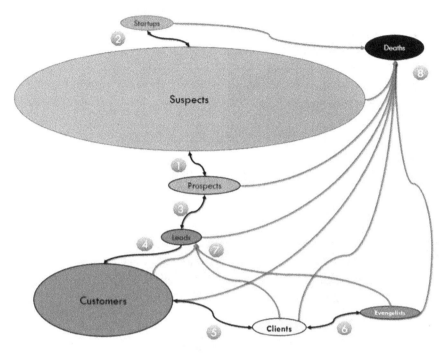

The sales channel is an ecosystem where entities flow from status to status through paths. Marketing is an optimization of the performance of these pathways. As the ecosystem runs, the size of the bubbles changes based on pathway performance. The customer bubble is empty at start-up, and your focus must be 100% on moving suspects to prospects and customers. As the system starts to flow, you must balance all the paths to available resources and business goals.

Terminology

To understand our Ecosystem chart, we need to explain our terminology.

> **Start-ups** are the beginning of the system, and growth occurs when start-ups exceed deaths. To march in place, a business has to have enough new business to replace the unstoppable deaths.

- ➢ **Suspects** are businesses that you suspect you can do business with.

- ➢ **Prospects** have shown an interest in your business. Transfer in this pathway is a Cold Call, First Contact, or Hello.

- ➢ **Leads** are when a customer or prospect shows an interest in your product or service.

- ➢ **Customers** have bought your product or service.

- ➢ **Clients** have a deep enough relationship that they will continue to do business with you, even if you make a mistake or are not the low-price provider.

- ➢ **Evangelists** are the source of the highly coveted Referral. A referral is when a person has a friend with a need you serve, and they recommend your business. Evangelists do not HAVE to be customers, but they are always "Trusted Advisers" of the prospect.

Transfer Paths

A line between one status represents a pathway. This pathway transfer is commonly called a conversion, and it involves some form of "Saying Yes to a sale proposal.

(1) Suspects become Prospects by showing an interest in what you offer.

(3) Prospects become Leads by considering a product or service pitch

(4) Leads become Customers by **Saying Yes** to a Sales Proposal

(5) Customers become Clients by continuing to do business with you even when you have made a mistake or are not the low-price provider

(6) Clients become Evangelists by referring a friend to your business.

Path Markers

On the Sales Ecosystem Map above, you will see number markers like the example above.

1. **Suspects to Prospects** happens when a business expresses interest in your product or service. Typically some form of Yes to an interest contact. We tell you that we sell X, and you say, "I am interested. Tell me more." That is a "Yes" response to a contact. Commonly called the Cold Call. These conversions are often the most expensive because of the high failure rate. PPC advertising is a common tool in this path.

2. **Start-ups to Suspects** happens when a new business is detected in the market. Detecting new businesses in a market is different from finding a new business. Many times, this conversion happens when an established business enters a new market for them. For example, as a Plumber business grows, it is common for them to enter the HVAC market or other home service markets. There are data sources for some of this, but the quality and timing are a challenge.

3. **Prospects to Leads** happens when the business expresses an interest in a specific product or service, and you start the proposal and negotiations phase of the sales relationship. There are return paths (7) on the chart, and these are the highly coveted referral sales. Everyone loves them, but there are never enough of them.

4. **Leads to Customers** happens when the deal is accepted and the order is processed. Many businesses make a big mistake and stop selling to the business at this point. In this conversion, competitive actions in the market, pricing, and past performance are a few of the driving forces your systems should watch.

5. **Customers to Clients** is all about developing a relationship that creates value in the mind of the customer. The deeper the relationship, the more it can change the results. Make a mistake or become a higher price provider with a customer, and they are gone. With a Client relationship, the business can survive the challenge. Your system should watch events and feed them to sales reps so they can use them to nurture this relationship.

6. **Clients to Evangelists** happens when a business tells their friends about your product or service when their friend "NEEDS" what you do. Evangelists commonly come from Clients or Customers but not always. The key here is that they are trusted advisers for the business with the need. Many times, CPAs, Lawyers, and other

trusted advisers are the source of referrals even when they are not customers.

7. **Back-flow from Customers**, Clients, and Evangelists is a specific challenge to Sales. The lines are the recurring orders from Customers and Clients and the highly coveted referrals from Evangelists. When they are a referral, the entity flow is from the suspects to prospect to lead.

8. **Deaths** can flow from any of the states, as shown by the lines. In every ecosystem, there is death, and it must be filtered out. In the body, the liver filters the blood and eliminates the waste, and this Road Map Marker is that process.

Size Matters

The Sales Ecosystem chart is a mature company. The size of each status helps us understand the scale of the sales opportunity distribution. For example, the first contact is by far the most expensive conversion in the system. Depending on the company, this can be 10x the cost of a referral or recurring order. However, without new customers flowing into the system, the business will slowly die. Attrition or death within the system is unstoppable. In a start-up, almost all the orders are new business, which crushes their profitability compared to a mature company. Some sales forces make the mistake of over-focusing on the more profitable sales and killing the business. A critical point that management must understand is the scale of these stages. It is only with that understanding of scale can they bring order and balance to the business.

Sales are a Process, NOT a Project

A process is infinite and has no end. A project has a defined start and end and can be completed. **Sales are a process.** Running a project without a defined completion is the original sin of project management. Sales are never done, and the conditions change continually. It is an infinite challenge worked on with a finite resource. In sales, **the grand challenge is "What's next?"** and there is always something next.

Managing a process requires a different mindset than managing a project. For a process, you must manage for the long term and accept that not everything is measurable, but everything is connected. Sales are unlike many other jobs where you can complete the task. In sales, as soon as you finish one, you must move on to the next. The key to sales optimization is deciding "What's Next?" To do this, let us start by looking at a typical market chart.

The secret is to continue to cycle customers into the lead-to-order process. The path from each status has its influencers. When dealing with new business, as we often are, the key is the path between suspects (those that you think you can do business with) to prospects (those that have indicated an interest in doing business with you). The prior sentence is complex, but It is important to understand.

Everyone struggles with the conversion from suspect to prospect. This pathway is the most expensive and least productive conversion but essential. New business costs more and produces less, but it is the source of Customers, Clients, and Evangelists. From those come the highly coveted referrals that cost less and produce more. The problem with the referral business is that it is impossible to scale. There is no silver bullet that will suddenly cause referrals to grow at the **rate YOU want.**

Hate New Business Dev? - Everyone Does!

Managing sales means managing the flow from status to status and regulating the size of each pool of business. Sales has a cycle of peaks and valleys created by the mismanagement of the system. The leads grow, and sales reps prioritize them above everything else. The first thing dropped is the cold calls because **everyone hates them**. This decision causes the new lead flow to stop. When the current leads are worked through, they go back to cold calls, but the pipeline is empty, and it takes time to get that reloaded. This common practice causes all sorts of operational issues, with demand growing and crashing over and over. With large sales forces, this averages out because different reps are in different parts of their cycle.

The challenge comes back to the infinite work and finite time. The best advice is to consistently work all the paths based on your best estimate of the finite labor resource. Then adjust until you find the right balance. Nobody is perfect at this, but some are better than others.

Now Think BIG

In the Sales Ecosystem chart, there are eight road markers. These represent the path flow between statuses. You should have a plan for how to manage each of those and understand the flow through them.

When you are new, you must put all your resources on path one because you have nothing else to work on. As the customer base grows and your referrals grow with it, the lead flow will come from the customers, clients, and evangelists. As this happens, you can transfer resources to the other pathways. You need to resist the temptation to cut new biz cold calls & advertising to zero. Your sales and profits will increase in the short term, but you will slowly but surely die.

Call to Action

Calling people to action is what Sales and Marketing are all about. The grand challenge is that getting people to act is more art than science, but you can depend on basics.

> You cannot predict the action of an individual, but you can predict the action of a group.

A simple rule of statistics is that the larger the group, the more predictable they become. When you design web content for a response, which is always, you need to use one or more of the items from the table below.

1. Trust	5. Consistency	9. Fear: Scarcity
2. Reciprocity	6. Social proof	10. Greed: Deals
3. Influencers	7. Authority	
4. Commitment	8. Liking	

Weapons of Influence

This section is influenced by Robert Cialdini's book titled Influence. We believe in giving credit to those that did the work, so here it is. His book is a great read, although longer than it needs to be. We are reducing his research to a few pages about how this specifically applies to you today.

Weapons of Influence contribute to an individual's propensity to be influenced, but they do not cause action. In simple terms, these are the tactics you can execute to get a reaction. When you examine your content, one or more of these should likely be in your content. If none of these are in your content, **you should be very worried**.

Reciprocity

People tend to return a favor. Reciprocity is created when you give the other person something of value without expecting something in return. The best form of reciprocity is created when the thing you give them is related to the problem you solve or the service you provide. The item of value can range from a flower in an airport (Hira Krishna) to a food sample in Costco or this book you are reading. We recommend meaningful content related to your business and the problems you solve. The reaction we sometimes see when we say this is that it would be driving business away by telling customers how to do it themselves. To some degree, this is true, but those people were never going to do business with you. The others will appreciate the complexity and work involved in your magic and seek you out to solve their problem.

Commitment and Consistency

People do not like to be self-contradictory. Once they commit to an idea or behavior, they are averse to changing their minds without good reason. Offer them a small step that is easy to execute, and the momentum of that decision might carry the deal to a close. Getting a small yes in the conversation with the prospect is important in this Weapon of Influence.

Social proof

People will be more open to things that they see others doing. The vehicle markings become a critical part of this social proof. When a person sees a neighbor using a specific business, and they need the same service. **Guess who gets called first?** Social proof comes in many forms, and a common one is public reviews. If you want reviews, you have to provide an experience worth talking about and ask for the review.

Authority

People will tend to obey authority figures. Businesses often overlook this one because they think they have no power over their customers. That does not mean that you cannot use this attribute in your marketing. You can seek out and do business with highly authoritative individuals and get some of this by association. Look to serve the powerful people in your community. Their authority can flow to your business.

Liking

People are more easily moved by people they like. People do business with those they like! Therefore, teaching the entire staff how to be friendly is good for business. This topic is so big that there are dozens of books on the topic. Start with basics like Active Listening and be an evangelist for this, and it will pay dividends for years.

> There are experts that propose that Fear and Greed are the only emotional states that cause a person to act.

Fear – Scarcity or Loss

A perceived limitation can of resources can generate demand. Fear is often related to scarcity, but there are other forms. Fear of Loss can be based on danger

Greed - Deal Seekers/Offers

Some people just need to think they are getting a deal. It seems to be in the DNA of some people that they must get a deal. For this crowd, coupons are a great answer, even if the deal is nothing special. Be careful not to train your prospects to put off purchase decisions, waiting for a better

deal. We have seen lots of businesses get into the deal spiral that creates a race to the bottom. Often, putting a deal in the market shifts the time of the purchase ahead and reduces the margin.

Trust

Some might put this under Liking, but Trust and Liking are vastly different attributes. The basis of the marketing through education strategy is based on this attribute. The concept is rather simple in that you take what customers are always asking and use that to drive content creation that will compete for searches under those questions.

> People do business with those they **trust**.
> They **trust** those that help them understand.

The bet here is that if they are asking your employees, they are also asking Google. If you create content to answer those questions, they will trust you. It is a short trip from trust to doing business with you.

The naysayers of this tactic say that answering the questions will create more DIYers and, to some degree, true. However, the person who has DIY in their soul will never be your customer, so nothing is lost.

Traffic Categories

Traffic can be categorized into broad types that help us develop strategies that work. To effectively create advertising or content for these types, you must get into the group mindset. As Yoda from Star Wars would put it, *"You must become one with the Audience."* Here are the major types that all SMBs need to consider.

1. Emergency
2. Service & Repair
3. Installation
4. Maintenance
5. Products
6. Impulse Buy - Deals & Promotions
7. Just Browsing – Display Network

Emergency

Emergency traffic is one of the first things that SMB's should go after. The traffic is highly responsive, and everyone knows it is making the traffic expensive. In many markets, it is right on the edge of not being profitable. The response rate commonly is 20% or better, so for every five clicks, the phone rings. We often run into Google Ad accounts below this 20% target because they did not pay attention to their negative keywords. Words like Plumber will drag in job and DIY traffic, just to name a few. Broad words in an emergency traffic campaign can set your credit card on fire and produce very little. You need to trim your traffic with negative keywords, or your cost per click will kill you.

Service & Repair

Service and Repair traffic is unlike emergency traffic because the consumer can and does shop the solution. Being first in this is not as important, although it is still a factor. The key to position in the search result is to make it to the shortlist that the consumer will reach out to and evaluate. This traffic will likely go through a quote and propose cycle because the consumer is not under any pressure to decide.

Installation

The installation traffic acts much like Service and Repair. However, the time pressure on the purchase decision is less. Once this type of lead gets into your sales pipeline, you must make sure that you have good follow-up processes in place.

Maintenance

Maintenance is or can be a major source of revenue for SMBs. Surprisingly many do not pursue maintenance opportunities. In most cases, all it takes is a reminder phone call or mailing to get the maintenance scheduled. If you install a product and know the life span is seven years, then **put that on a follow-up calendar**. If you coordinate this with your promotional calendar, then you can rule the market.

Products

Like all industries, new products are rolling out all the time. Every time that happens, it creates a new opportunity to reach out and create more new business. Most products put you at the front of the line to get the maintenance work that commonly comes with the product. There should be an original blog article and another distribution cycle of your marketing for every new product.

Deals & Promotions

We consider deals and promotions to be the last tool in our marketing toolbox, and for a good reason. If you win business as a low-price provider, you will also lose it for the same reason. People attracted by a deal are not loyal customers and will grind you on every penny. Exciting deals and promotions are necessary for your marketing plan but make sure you understand the long-term cost, not just the immediate effect. If your annual marketing calendar has more than three deals or promotions, rethink your calendar.

Resource Strategy

In a Word War, there is a truth that nobody likes to hear, but everyone has. **Goals are infinite; resources are finite.** You must be smart about where you put your resources because you will run out.

For your strategy, you must seek balance. In most cases, this means limiting the market you reach to, the aggressiveness of that reach, or the game you play. They always separate the AO (Area of Operation) into secure and not secure in military operations. The secure area is called the green zone. For marketing, it is the most prime placements or keyword targeting. Outside the green zone, the investment in security resources is much more limited.

The second part of this is the game you choose to play. We will shift away from the Word War storyline for a second here and over to sports. Your marketing can be played like football or basketball. In football, the competition is consistent from yard one to the touchdown. Every inch is fought over. In basketball, most teams shift tactics when the ball is past half court. Before that, the competition is light and more of a relaxing time followed by high intensity as they approach the goal.

Marketing during slower periods has this same decision. In many Home Services, like plumbing, business slows down during the holiday season, starting with Thanksgiving and ends with New Years'. The slowdown is because of planned events that put off routine services. Emergency work does not drop, but work can be scheduled and controlled does. So the question is, what do you do in this period?

Common strategy during this time is either to back off and reduce your marketing expenses - the basketball strategy. Or run at full speed - the football strategy. Both are perfectly valid, and each has its advantages. The basketball strategy saves your budget for a more competitive time. The football strategy picks up lower-cost work because competitors have a basketball strategy and search is a market demand-driven system.

Stop Wasting 50% of your Ad Dollars

"Half the money I spend on advertising is wasted; the trouble is I don't know which half."

John Wanamaker
Father of Modern Advertising

Learn how to fix this

John Wanamaker is a legend in marketing, and in his day, finding wasted advertising was impossible. The world has changed, and the good news is that his statement is no longer entirely true.

Today, we can measure things that were impossible in Mr. Wanamaker's day, and this is especially true in digital advertising. Google Analytics gives us a measurement that digs deep into this problem, and **it's called the Bounce Rate.**

Bounce Rate is the percentage of traffic that loads one page and then leaves your site with no other interaction. When traffic acts like this, then it is a reasonable argument that the advertising investment was wasted. Reasonable YES, correct NO.

A major flaw exists in the way this measurement is taken. Here is a bounce scenario where the bounce is NOT wasted traffic. A person goes to your site, sees your phone number, picks up the phone, and calls you. Most businesses consider this a home run and the reason you advertise, but it is a page visit that Google considers a bounce.

Let me explain the tracking problem. Google Analytics gets the starting time-on-page from the page load, but end-time comes from the load of the next page. In the case of a bounce, there is a start time with no end time. To solve this problem, have a script wait a reasonable period and then fires off an event to Google Analytics. The page gets an end time and breaks the bounce. In WordPress, this is as easy as adding a plug-in to your page template or JavaScript into Google Tag Manager.

This un-bounce function must have a wait time after the page load and before creating the event. Typically, we set this to 10 seconds, but that is a judgment call based on the type of business and the website design. The time must be long enough to get your real bounces because that is what you must work on. My argument for 10 seconds is that it is long enough that the person is probably engaged with your copy. Under 10 seconds gets your real bounces reported.

Now that you have your real bounce rate. **What do you do with it?** Great question. To start, examine the traffic source that created the bounces and be careful of chasing outlier data. Everyone wants all visitors, especially those from paid sources, to take their planned action. The reality is that not everyone is going to do what you want to be done. Some level of bounced traffic is unavoidable. The question is, what is acceptable for your business?

In Search, drill down to the keyword to get clues of where this is coming from. The tools you must improve this includes the keyword, search terms, and ad copy. The search terms are a bit of an enigma because you know the keyword bounced but not which search term bounced. Negative keywords are a powerful tool; however, you must be careful because they can be a silent killer of your account.

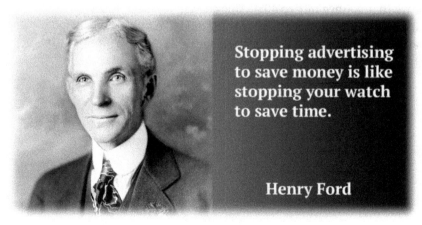

Stopping advertising to save money is like stopping your watch to save time.

Henry Ford

Ad copy might surprise some, but it can be a highly effective tool. Most of the time, you're thinking about ad copy is **"How do we get the**

Click?" but another way to think about copy is as a traffic filter to stop unwanted and poor-quality clicks from happening. We had a client that did software development, and their projects started at $10,000 and went up from there. When another agency first wrote the ads, they were tuned to get the click, and they did. The result was lots of traffic, but the quality suffered. With a good understanding of what the client needed, we changed the ads to include the project size. The number of calls dropped, and the quality of calls improved. The client was pleased with the change in calls. The client got fewer calls, but they were the right calls.

When you tighten the keywords, it drives volume down and quality up. Follow the math, and this will improve the conversion rate, the ultimate goal. What you do not want to do in this process is reduce the budget to take the savings. In marketing, it is easy to fall into the **"Saving yourself out-of-business problem."**

In the Google Display Network, the process involves a higher degree of geek magic. The targeting accuracy in Display is much lower than in search. It often requires designing the ad groups to separate pools of traffic. There is more magic and less math, but you can squeeze improvement out of the display network using direct placement targeting.

Suppose you have a $1,000 budget with a 50% bounce rate and improve that bounce rate to 40%. This change will save you $100 a month, which is $1,200 per year. If you would like to **stop wasting advertising dollars,** give us a call, and let's talk.

Order of Battle

Just like in the Military, Digital Marketing has a common **"Order of Battle."** The roll-out sequence below is the most common, but this is not universal. The roll-out sequence is important because some channels have prerequisites. For example, trying to roll out SEO without the data from PPC is a risky decision. A business can roll some of these out of order, but that increases the risks. Here is the standard order:

1. Website
2. Tracking
3. PPC (Pay Per Click)
4. SEO (Search Engine Optimization)
5. Social Media
6. Email

Website

The Website is the first part to be rolled out because it is the foundation and the hub of your Word War. The website sets the keyword strategy, establishes messaging, and sets brand standards. A good website design will serve your business for 3-5 years or more.

Tracking

Tracking is what separates the winners from the others in Digital Marketing. If you do not know what is going on within your website, then **you are flying blind**. Designing your goals and sub-goals is critically important. You want to track everything early and often. It is not enough to track orders or leads. You must track the earlier steps that lead to this. For example, if you track an inbound call as your conversion. Then track the steps that got them there. In most cases, a lead has multiple touches.

Many businesses use a "Marketing Through Education" strategy. When doing this, you want to track the engaged reader as an early-stage conversion. In this context, an engaged reader is a visit over a certain time on-site. Depending on the length of the content, this can be 2 minutes or more or multiple pages. Over time you will discover that there is a relationship between engaged readers and calls. These engaged readers allow you to see a higher volume of conversions allowing you to see winners earlier.

PPC – Pay Per Click

Many businesses will look at PPC as an expense to be avoided, but it is not. It is a necessary tool on the Digital Marketing Utility Belt. PPC expands your knowledge of your specific market and the value that it puts on your traffic. When you get to SEO, this will be critical information needed to make some tough calls. Google Analytics will tell you what is happening on your website. Google Ads will tell you what is going on with competitors and the market in general. If understanding your market share is important, PPC is part of that formula.

The other reason that PPC goes first is that it responds faster than the other channels. It is reasonable to set up a Google Ads account and have traffic flowing to your website in under 24 hours. **It is not free, but it is fast.**

SEO – Search Engine Optimization

SEO is a complicated, rapidly changing, brutally competitive field in Digital Marketing. Everyone wants it to be **fast, free, and easy**. But it is none of those things. Winning at SEO takes time, dedication, and smarts. Google made more than 3,000 changes last year, and each one of those is either a win, loss, or disaster. SEO is a process, not a project, and it changes with every search. The fundamentals have not changed for years, but the finer points change constantly.

Social Media

Unless you have been living under a rock for the last decade, you know that social media has changed the world. Business Leaders have a tough time understanding Social Media because they want it to be free advertising. They want to get up and deliver a speech and have people react the way they want them to. Social Media does not work like that. Social Media is a cocktail party, not a rally with a speaker. The strength of your voice in Social Media is a product of your network of connections. Building and maintaining your network takes time and effort. The timeline for building your social media network is measured in years, not days.

Email

Email is one of the most powerful tools on your utility belt, but it is a long, slow, hard battle to build and maintain a list. We are sure that someone has already pitched a list they can sell you, but that is not the same. That is SPAM, and it creates more problems than it solves. To build a list, you must exchange some value with the recipient in exchange for permission to email them. If you buy from a List Broker, what you are buying is not an opt-in list. If the recipient signed up at all, they did so for some value. Unless that value came from you, it is a SPAM list.

To keep your list in decent shape, you must communicate regularly without becoming a pest. To do this, you need a steady flow of valuable content, and that is no easy task. When you launch your email list, you will need to add a blog to your site for the ongoing content. The blog will help your content strategy in PPC, SEO, and Social Media.

What's Next – Rethink and Refine

Sales is a process, not a project. The next step after everything is working is to find what's next. Our advice is to look for the lowest performer and find ways to move that up but not too much. Systems work better if you bring all the parts up in sync with each other. Small incremental increases are what you get as the business matures. Breakaway performance is what you get from innovation, and innovation is hard to replicate.

Customer Modeling

The foundation of a Sales Strategy is an understanding of the Ideal Customer. You cannot target or communicate with a group that you don't first define. This concept is approached differently by different experts, but ultimately your ability to communicate is based on:

- Who are they?
- What they value?
- Where are they?
- Why do they care?

We purposely skipped over **"When"** because in Digital Sales, when is always the *present*. To model your ideal customer, we propose that you must consider:

- Demographic Data
- Firmographic Data
- Personality Type
- Business Role
- Generation
- Audience Share

Demographic **information** includes age, race, ethnicity, gender, marital status, home ownership, income, education, employment, and more.

Firmographics are to businesses what Demographics are to people. Firmographic information includes Industry SIC, Employees, Sales, Credit Rating, Website, Location, Growth Indicator, and much more depending on the source.

Personality Types give you an understanding of how this person receives and processes information. No one person is completely one personality type, which can vary depending on the communication circumstances. One of the common models is Dominance, Influence, Steadiness, Compliance (DISC).

Business Roles often correlate to personality, but it is a soft relationship. Roles in a business are Technicians, Managers, Rainmakers, Entrepreneurs, and Worker Bees. Roles drive how the person finds value in your product or service. What is positive to one could be a negative to another.

Generations give life experience references that can help you communicate with the group. How communication is received and processed often varies by the Generations. For example, it makes no sense to sell mobility or assisted living to Gen Z, but it certainly fits the Silent and G.I. Generation. Although, a Millennial might be a key part of the purchase decision for these items. Generations give you common life experiences you can tap to help communicate with that generation. For example, for Baby Boomers, Vietnam was a life experience often with strong emotions. For Millennials, Vietnam is a history lesson with little or no emotional connection. Emotional connection is how many sales happen.

Audience Share indicates what this profile represents as a share of the audience. When building out your ideal customer profiles, you want to stop somewhere between 80-100%. To help teams remember the profiles, a best practice is to give profiles common names and photos that visually communicate the profile.

With your audience profile documented, you can start to examine messaging that talks to that profile. There are always exceptions to every statement made any time you deal with this type of data. Stereotypes get a bad name because they do not fit perfectly, and they create social problems. However, they are often fully accurate for 90% of the group.

Data Quality

Modeling requires a degree of guesswork, but you should strive to make this as accurate as possible. For existing businesses, this means looking at your best current customers. What is the gender mix? What generation do they belong to? Guess where you must but do your research. It will pay big dividends.

B2B Model

Below is an example model. The photos are not essential, but over the years, we have found that giving them a picture and a name makes it easier to discuss in strategy meetings. We can talk about how Bill or Betty might take specific messaging—putting these in front of your team when building content can have a huge effect.

SMS   www.smsrd.com   (800) 272-0887

Ideal Customer Profile

Description	Small Business Owner
Audience Share	75%
Geographic Target	United States
Target Growth +/-	20%

Demographics

		80% Low	80% High			
Age	50	35	65			
Male	82%			Bill		
Female	18%			Betty		
Minority Owned	18%					
Veteran Owned	8%					
Married	70%					
Income	$80,000	$80,000	$150,000			
Home Owner	90%					
Touch	5					
Pipeline Days	45					
Generation	G.I.	Silent	Boomer	Gen X	Millennial	Gen Z
	0%	5%	60%	30%	15%	0%
Personality	Decision Maker	Encourager	Facilitator	Tracker		
	60%	10%	15%	25%		
Role	Technician	Manager	Rainmaker	Entrepreneur	Worker Bee	
	70%	10%	10%	10%	0%	

Firmographics

	<20	21-50	51-100	100-250	250-500	>500
Employees						
	90.00%	5.00%	2.00%	1.00%	1.00%	1.00%
Revenue (millions)	<1	1-3	4-10	11-50	50-200	200+
	90.00%	5.00%	2.00%	1.00%	1.00%	1.00%
Industries	Pro Serv	Home Serv	Mfg	Distributor	Other	
	35.00%	40.00%	10.00%	10.00%	5.00%	
Square Footage	<1500	1,500-2,500	2,500-5,000	5,000-9,999	10k-20k	>20K
	75.00%	25.00%	10.00%	10.00%	4.00%	1.00%
Website	Yes					
Headquarters	Yes					
Franchise	No					

B2C Model

Business to Consumer (B2C) and Direct to Consumer (D2C) models have different models because the firmographics are not relevant. Here is an example of what this model would look like. In almost all cases, you would extend this with something about the consumer. In this case, it is connected to Homeowners, but it could be any of the common consumer segments. Things like outdoor enthusiasts and many others will make your models better.

SMS	www.smsrd.com	(800) 272-0887

Ideal Customer Profile

Description	Home Owner
Audience Share	75%
Geographic Target	United States
Target Growth +/-	4%

Demographics

		80% Low	80% High			
Age	50	35	65			
Male	49%		Bill		Betty	
Female	51%					
Minority Owned	18%					
Veteran Owned	8%					
Married	82%					
Income	$65,000	$40,000	$120,000			
Home Owner	100%					
Touch	5					
Pipeline Days	45					
Generation	G.I	Silent	Boomer	Gen X	Millennial	Gen Z
	1%	5%	48%	30%	15%	1%
Personality	Decision Maker	Encourager	Facilitator	Tracker		
	50%	10%	15%	25%		
Role	Techician	Manager	Rainmaker	Entrepreneur	Worker Bee	
	70%	10%	10%	10%	0%	

D2C Manufacturers

Direct to Consumer (D2C) Manufacturers are continuing to grow in the market. This transformation is just one of the many business model adaptations driven by the Internet. In the olden day's manufacturers supplied distributors, who supplied retailers, who sold to consumers - **the world has changed.**

Do you HATE Amazon?

Well, you are not alone! The Internet created a class of Super Retailers, with the biggest, ugliest, and nastiest of all being Amazon. Amazon has business practices that are on the fringe of antitrust laws. They are creating private label products under Amazon Basics and dozens of others. These private label brands are a competitive threat to manufacturers that reach a successful level on Amazon. Build the market, and **Amazon might be your worst competitor.**

In just a few years, Amazon ripped nearly a third of the market for batteries from Energizer and Duracell. This at a time when they were selling competitive products. It must be nice to examine the sales volume of your competitors down to the end consumer. The New York Times in 2018 estimated that in-house Amazon brands could break $25 billion in sales soon. Success in Amazon could result in Amazon sniffing around your market.

The Opportunity

Selling directly to the consumer allows your business to pick up the distributor and retailer margins, but it comes with additional responsibilities. In the classic distribution model, the Manufacturer was responsible for the brand advertising, and the retailer did the sales advertising. If you open the D2C channel, you need to allocate dollars to replace what the retailers were spending.

Digital Advertising has more control of the delivery target, which can be hugely beneficial to the D2C channel. Consider your national strategy, excluding those areas that strong retailers represent. This way, your good retailers do not see you as a competitor in their local placements.

The Risks

Amazon has too much power and can exercise undue influence on manufacturer's pricing and margins. We have seen Amazon act without regard to the impact on other businesses. When they own the path to your customer, that is a risk you must mitigate. The only way to mitigate this risk is to own the path to the customer. **That means selling direct.**

If your sales depend on Retailers, they can react poorly to a manufacturer who tries to take them out. To change from a manufacturer to a competitor can spoil even the best business relationship. Most retailers have more than one source for common problems, so you must be concerned with them shifting sales to your competitors.

All sales are local, and retailers know the local community much better than you ever will. The odds are good that they will know the best way to reach the local community. They have the advantage of seeing the whites of the eyes of the customers. They know the localized objections and how to answer these. Therefore, Manufacturers have had co-op advertising funds for generations.

MSRP & MAP

MSRP = Manufacturers Suggested Retail Price

MAP = Minimum Advertised Price

These policies keep the market chaos to a minimum. Without these policies, a price war can spiral out of control. You are setting the stage for a pricing war that drives margins and perceived value down. It is good policy to have this in place and even better to ensure that your business complies with the rules. **It can be catastrophic** to have a retailer discover they lost a sale to you and the customer got the product at a price below your MSRP and MAP. If you think for one second that your price quote will not get shared with a local retailer, you are sadly mistaken. There is no such thing as a secret in sales.

The Bottom Line

D2C is an opportunity that all manufacturers should explore. However, like all marketing and business relationships, it can be complicated. Weigh the risks and rewards and test, test, test.

Study Industry Leaders

This section is not about studying your competitors. It is about your industry leaders. If you are unfortunate enough to have one of these in your market, then read this section twice because your path is especially dangerous and difficult.

We picked Plumbers as our example industry, but this same process applies to all markets. Study the leaders first and compete with them first. This way, you train for a level much higher than your market

You can learn a lot about an industry by studying the businesses that are killing it in the market. One measure of business growth is the Inc 5000 list. This list details the top 5,000 fastest-growing privately held companies in the US, and the competition at this level is fierce.

To give you an idea of these processes, we picked the Plumbing industry. In 2019 the Plumbing Industry had 15 companies that made it to the Inc 5000 list. According to the latest data from IBIS World, there are 120,749 plumbing businesses in the US. These companies represent the **top .01%** of the industry.

To kick this off, we downloaded the top 10 of these. The sites will probably be different by the time you read this book, but here are the website names that are unlikely to change:

These companies are worth studying

1. https://www.jdprecisionplumbing.com/
2. https://easyasabt.com/
3. https://milltownplumbing.com/
4. https://flowrightphi.com/
5. https://www.callschaalyaall.com/
6. https://www.chesapeakeplumbingandheating.com/
7. https://www.aaplumbingsa.com/
8. https://866411zapp.com/
9. https://www.sunshineplumbingheating.com/
10. https://tomdrexlerplumbing.com/

Here are the thumbnails for these industry leaders:

https://www.callschaalyaall.com/	https://chesapeakeplumbingandheating.com
https://866411zapp.com/	https://www.sunshineplumbingheating.com
https://tomdrexlerplumbing.com/	https://www.aaplumbingsa.com/

Observation Summary

1. Blue is the dominant color.
2. Common Contrast color is red and orange.
3. Pages are highly graphic.
4. All sites are professionally designed.
5. 7 of 10 have a staffed Chat Function.
6. 8 of 10 have dominant phone numbers.
7. All sites have tracking installed.

Granted, this is a very shallow data dive into a very deep pool, but studying the top of your market can teach you many things. From these ten sites, we built a market composite word cloud. A word cloud is simply a way to visualize the text on the collection of pages.

So, what does this data say to you? There are many interesting things in this data, and reading data is an art. Word Clouds are a great tool for visualizing the message in the text. In this cloud, the size of the word is the occurrences of the word weighted by position because hierarchical position matters. When We first looked at this, it said several things to me. The most glaring is that these are Plumber sites, and Plumbing is the dominant word. These are smart people, so they know that "Plumber" plus "City Name" is the money search. Yet, they are optimized for Plumbing – Ouch!!! If a person is looking for plumbing, it is more likely a DIY (Do it yourself) search. Not the service search the plumber wants. Plumber and Plumbers show up in the Top 50 words but with a weight of 4.5 and 6.6 compared to 31.3 for Plumbing.

This model includes 1,341 words after some minor cleanup. So, the Top 50 represents the top 3.7% making each of these words especially important. Any respectable SEO Expert will tell you that to waste words in your top 4% is a crime of epic proportions. At first glance, We can see several words. We never want to see this type of word cloud. Words like: 'and,' 'to,' 'the,' 'and' are a waste of SEO juice.

One interesting tidbit here is that Plumber did not make it to this list. It is in the data but at a lower level. If a site is optimized for "Plumber" and "City Name," we should find it at the top of this list.

All the Top 50 targeted words are in the cloud, with contributions from the prime positions of Title, Meta, and Headers. The placement of common words in high-ranking positions is an optimization error. In this case, we believe they opted to give preference to the web experience delivering more sentences than statements in these prime positions.

Taking Words Out of Optimization

Sometimes the web experience needs to keep a common word in a prominent position, and there is a way to fix this. This tactic applies only to **the most competitive situations**, but you can change the word into a graphic and then code that picture with the target word. The tactic removes the word from optimization and makes it human-readable only. Again, this is only for extremely competitive situations. In most cases, you simply remove these from the high-value position.

The most common words across these sites are:

word	Count	Wgt	Title	Meta	H1	H2	H3-6
plumbing	141	31.3	5	16	2	6	1
and	250	31.3	4	13	3	4	2
the	216	24.3	2	7	5	4	1
to	192	20.2	0	6	0	7	5
services	86	17.3	2	8	2	2	2
our	121	16.9	0	5	4	4	4
in	96	15.9	2	7	2	2	7
service	87	15.6	0	7	0	8	5
air	67	14.4	2	5	1	3	2
your	114	14.1	0	7	3	2	4
a	113	14.1	1	9	0	2	1
us	91	13.1	0	0	1	2	3
for	76	12.3	0	8	2	3	3
water	77	11.7	0	0	0	1	0
heating	54	11.6	1	6	1	1	2
you	129	11.3	0	0	1	7	0
repair	75	10.1	0	0	0	0	0
call	45	10.0	0	8	0	5	2
of	83	9.2	0	5	0	1	2
conditioning	38	9.2	2	3	1	1	1
is	82	8.9	0	5	0	1	1
hvac	37	8.7	3	3	0	3	0
drain	35	8.0	1	2	2	0	2
we	96	7.8	0	3	0	1	1
more	45	7.3	0	2	0	0	1
plumbers	17	6.6	2	4	1	2	0
cleaning	32	6.4	1	0	2	0	2

The Grunt Test

In military terms, a grunt is an Infantryman or Marine. They are the tip of the spear that faces the first contact with the enemy. In marketing terms, this is the first contact. We would love to take credit for this one, but we must give credit to Donald Miller from Storybrand. The concept is simple. In a glance at your page can a person:

1. Understand what you offer?
2. How it applies to them?
3. What do they need to do to buy?

If a new visitor cannot answer yes to these three points in the first few milliseconds, then you are not done with the page. In businesses, four common perspectives come to the page design. Sometimes these are all assembled under the theory that more-is-better, and we get a design committee. Here is what happens:

Technicians, Engineers, Scientists, Manufacturers, and CTO's cannot resist the temptation to start with three paragraphs demonstrating how smart they are and all the cool features they built into the product. These are followed by the technical product specs that go on for pages.

Clue 1: Customers don't care.

Managers, Accountants, and CFO's cannot resist the temptation to spend three paragraphs making sure the customer understands the terms of the purchase. They are followed by a long legal document with all the terms and conditions of the business relationship.

Clue 2: Customers don't care.

Entrepreneurs and CEOs' cannot resist the temptation to spend three paragraphs explaining the vision and the next version. They are followed by a vision statement and a development roadmap for 3-5 years.

Clue 3: Customers don't care.

Sales cannot resist the temptation to answer all the common customer objections. They know the questions they will ask, and they are driven to answer them.

Clue 4: Customers don't care.

The committee agrees that all of these are important points. They build a website specification requiring that all the points above. They develop a checklist for the designer and negotiate a development agreement.

Clue 5: Customers don't care.

The customer still does not understand what you offer, how it applies to them, and what they need to do to buy - and that they do care about.

Here is a clue – They DON'T!!!

Now look at your page and see if it contains any of these landing page sins. Answer these three questions, and the market will reward you. We know that the political realities will creep into our conversation because these stakeholders are power players in the business. Give them their points; make sure it is not on the first contact, which is the top of the page they land on. Add a learn more button or add more detail under the top page section but get these three questions answered with a glance at your page.

Let us get specific with this. A typical reader can digest 200 words a minute, and a glance is less than that. You need to use less than 50 words to answer these three questions. The total space taken for this job cannot exceed the top of the first page. If the person must scroll the page down, **then you failed.**

Setting the Marketing Budget

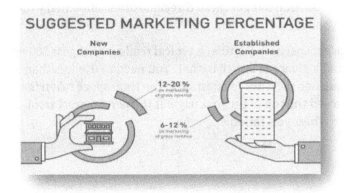

SUGGESTED MARKETING PERCENTAGE

New Companies
Established Companies

12-20 %
on marketing
of gross revenue

6-12 %
on marketing
of gross revenue

From 1% to 20% of sales

Marketing budgets as a percentage of sales vary by industry, competition, and media. In services, 6-20% is common, with product-based sales being lower than services. If you have a good accountant, they will have access to industry averages to get you closer to reality. Let us discuss some of the big variables.

New Companies must spend more than established companies. Unfair, but it is the reality of the hard-cruel cold market. Established companies have built their brand, and people know who they are. It takes less investment to cruise than it does to accelerate. This reality is as true in your car as in your marketing. When you enter the freeway, you must push the gas much harder than the cars already up to speed. Marketing has the same challenge for the same reason.

If you are in your first few years, a marketing budget of 20% is not excessive, but it needs to perform as an investment, not an expense. If your marketing is an expense, you are in big trouble, and you should call your mommy because success is unlikely, and you will need your mommy's emotional support.

Assuming the investment is profitable and having the resources to execute the sale, there should be no budget. The core of your marketing strategy is your CPA (Cost Per Acquisition or Action) and VOA (Volume of Acquisition or Action). If the CPA is being met and the VOA is still short, then keep spending.

> # "Spend as much as possible but no more than is necessary."

This rule does not mean that you should spend money you do not need to. As your business matures, the spend should come down for the same Volume of Leads. The ad-spend should only go up when you are adjusting the VOA for an updated sales goal. The key here is to think and manage this as an investment, not an expense.

Cost Per Action (CPA)

CPA is an acronym for Cost Per Action and is the cost goal for an

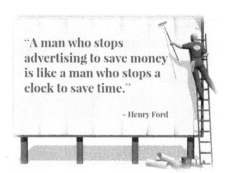

"A man who stops advertising to save money is like a man who stops a clock to save time."

- Henry Ford

average transaction. Later in the book, we will discuss how you get to this number. You need to understand that this is the cost goal, at which point the business goals are being met. The CPA is the trigger data point where you stop and build a corrective plan to bring it back under the CPA goal.

Every time we have the first CPA talk with a client, they all tell us that the lower the CPA, the better. Some have told us their goal was a CPA of zero. This goal is tactically smart and strategically dumb. Getting to a low CPA is easy. You take the cream of the crop, and most of the time, this is your branded traffic. The challenge with this is that the volume is not high enough to reach the business plan sales goal. Taken to the extreme, you can get to zero by turning your advertising off. You will effectively save yourself out-of-business.

Volume of Actions (VOA)

VOA is an acronym for Volume of Actions. VOA is the volume goal for the accounting period, typically a month. We recommend setting a target, minimum, maximum, and actual VOA. Minimum is what you need to operate the business to the plan level. The Maximum is the level at which you shift to saving your advertising dollars. The Maximum CPA is typically an operational or resource limit. Examples of a Maximum VOA are when all resources are busy, and the next step is a major investment. If you are renting Hotel rooms and the Hotel is fully booked, then you hit the Max VOA - **stop spending your ad dollars**. If you are a Plumber and run out of trucks - **stop spending your ad dollars**.

Competitors Set the Market Price

Because Google Ads, Microsoft Advertising, and all their cousins are market-driven systems, your competitors drive the CPA. Just like in an auction, they are not going to roll over and just let you have the leads at the lowest cost.

Budget = CPA * VOA

With the CPA and VOA set, the budget is simply CPA times VOA.

New Businesses

Most new businesses have a shortage of cash, and this drives their VOA strategy and roll-out planning. Like all businesses, you must spend money to make money. Money comes in two forms: creativity and cash. Make sure you spend the creativity first! Every market and campaign are different, and with that comes risk. Our recommendation for new businesses is to start slow and make your mistakes with the tap only partially open. You will find that marketing is a bit like turning on a faucet in a new house the first time. It tends to spit and sputter until it gets the air out of the line.

Minimum 4 Month Roll Up Plan

There is significant risk in a new account, so start slow. We recommend that you start at 25% of the final budget and then step up 25% each cycle to meet your CPA goal. If you fail to get your CPA, then either stay at that level and fix the problem or step back to the last step that did work. We understand the desire to get to the results you want immediately, but a new account is dumb and blind. Please wait for the data, think about what it means, and then act.

Secrets to Maps

For many customers, the business location is a big consideration regardless if they travel to the business or not. If the business travels to them, customers will think that the travel is at their expense. Either way, location is a big deal, and map positions are extremely competitive.

Maps are a hybrid position meaning that they combine paid and organic in the results. Unlike organic keyword pages, map listings compete on citations, keywords, reviews, and the reviews' authority. There are only three map listings on the first page, and there are always more than three eligible businesses. Who earns those coveted three positions depends on the reviews and the reviewer's authority? Let us take this apart and examine how each of these works.

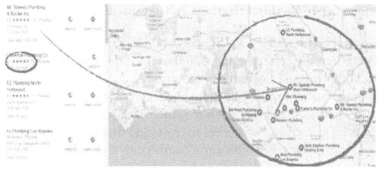

Citations are the first level of competition. Google is looking to verify that the SMB is at the address they claim. Google will look at the website and other sites to see if they can trust the address. The more times it finds the same address, the more it will trust the data, but it stops counting at a certain point. Five citations from good authority sites are just as good as a hundred because once Google accepts the location as valid, there is no extra credit.

Keywords are checked to see how the site ranks for the search. For this, you can refer to the SEO section that discusses the Keyword Organic Results. While the site's authority will impact the result, you can think about this as a yes or no eligibility test.

Step 1 – Claim Your Listing

To start, you need to claim your listing and get your contact information accurate. This listing claim only gets you into the game, and it does not get you on the SERP. The highly coveted first-page listing goes to those that win the reviews. Never let someone outside your business own the listing.

Citations

Citations are other sites that point to your site that include your physical address. The best examples of this would be your listing on the Better Business Bureau or your Local Chamber of Commerce. While these are the common ones, they are far from the only ones you should be concerned with. If you belong to a trade group that links to its members, make sure your data is up to date and includes your address. Almost every community has directory sites, and you want to make sure your business is in these and have your correct physical address. If you carry specific brands, make sure you are on the manufacturers' dealer locator as these **are high-value citations** and SEO backlinks.

Citations: Trust & Authority

Not all citations are created equal, and the citation's value is relative to the trust, authority, and relevance of the site that is pointing to yours. The higher the website's profile, the more juice it gives to the citation, and this is the same as the SEO process of simple backlinks. Google is surprisingly good at knowing what sites to trust, and these are the same places people trust offline. The BBB and Chambers are examples of this because they are highly trusted sites.

The Most Important Citation

Your website is the first, simplest, and most important citation. We have reviewed thousands of SMB websites, and it amazes us the number of times we find the business's website missing the physical address. Your website is connected directly to your listing, making perfect sense that Google would want an address citation there. Your address should be in text format on the bottom of every page, along with your phone number. We realize that nobody shows up at some business but finding the address is key to getting on the map. There is little downside to having your address on your home page, but clearly, it counts for maps.

Reviews

Reviews count, and you must ask for them. When you have completed a job, and the customer is happy with the results. Please give them your business card and ask them to for a review. Failure to ask almost always results in not getting a review. On your business card, make sure you include instructions on how to enter a review. There are thousands of places on the web to leave reviews, but we recommend that you focus on just a few; Google and Yelp.

Reviews are powerful, but power can swing both ways. Reviews can and will go negative, and you must be ready for that. The first rule is to keep your cool and respond to the negative review. If you know who posted the negative review, call them, and resolve the problem. Suppose you do not know the person posting the negative review. Then post a public response offering to make it right. Never respond in anger and never get into a fight on a review site. There are times when the best advice is to turn the other cheek. No business ever won a flame war on a review site.

Conflicts

It only takes a few minutes for most businesses to realize that they want to appear in cities that they do not have physical offices in, which is one of the great challenges. If your business serves more than one city, you will want to appear in all those cities. The problem with this is that these positions are competitive, and competitors in those cities typically consume up all the available slots on the first page. You can, and should, define your service area in Google, but that will not get you into competitive slots. The problem is that the more you are about one place, the less you are about others. You must decide what your primary city is and compete for that. Use your PPC ads to get your site to the secondary cities and make sure the ad copy is localized.

Gaming the System

With the "trick of the day," you might create a short-term win, but Google is brutal when they find someone gaming the system. The normal response from Google on this is to rip the entire site out of the index or put it into penalty status. If you think that they will not do this to you – you are mistaken. They have done this to very high-profile accounts, including little businesses like BMW, JCPenney, and even Google! One trick that Google is very aware of is the creation of ghost offices in other cities. The game is to find an address to use in that location and claim it as your office. We can guarantee you that you will get caught at this. Google will call and mail to the location to validate it. Your competitors will likely complain, creating an investigation by Google.

Summary

Your business must be on the first page of Google, which means competing in Google My Business and Google Ads. Use Places when you can get it and Google Ads when you cannot. Think of these as distinct types of warriors like Infantry and Armor. You do not want to go into the fight with just one and do not want your marketing with only one skill.

Test or Die

Never Stop Testing

THE ROAD TO
SUCCESS
IS ALWAYS
UNDER
CONSTRUCTION

Changing your mind is a sign of intelligence

We often ask people, *"What does it feel like to be wrong?"* They respond with words like horrible, crushing, and other derogatory terms. We then pause and tell them that we appreciate their feelings, but this is the answer to a different question. They are answering, *"What does it feel like to **discover** you are wrong?"* The simple truth here is that nobody likes being wrong, but you are wrong when it feels like you're right. In marketing, you must experiment to learn, and with that comes the very real risk of being wrong, and that is okay. If you are never wrong, then you are not pushing your thinking far enough to **find breakthrough ideas**.

The issue of Experimentation is a matter of time and degree. When you are barely getting by, experimentation is risky, and it's best to stick to the basics that you know work. As your business evolves and you build reserves, you must start experimenting because that is **where competitive advantage hides**. Everyone knows the sure things, which makes the competition in those just a decision of money, as in how much you spend.

For SMBs, PPC is a good example of this. Everyone knows that when managed properly, PPC works. However, it is also expensive for a known performer. Throwing money at a known performer will push your profitability right into the ground.

As you can afford the risk, you need to move more budget to experiments. Some of these will fail, but some will not. The simple rule here is to feed the experiments that feed you and kill the others. Some tactics require time to know if they are working, and you should not experiment if you do not have the resources or patience to see this through.

Read this in your best Yoda (Star Wars) voice:

Succeed or learn, there is no fail

Innovation or Incremental?

Experimentation has two different forms, and they come from innovation or incremental strategies. When you set up an experiment in your content or advertising, you must know what type you are working with. Innovation is riskier but has the potential for breakaway performance. Incremental improvement is safer but will only give you a small improvement. If your goal is to improve 20%, then go incremental. Over 50%, and you are solidly in high-risk innovation territory.

Incremental: Talk to your customers, sales reps, study competitors, and look for new leverage. Test shifting messages, new headlines, stronger calls to action, and other experimental ideas.

Innovation: This must come from you as you stretch your creativity. The

"If I had asked my customers what they wanted I would be raising faster horses."
Henry Ford

last thing you want to do is talk to customers, sales reps, or study competitors because that will give you incremental improvement. How to Improve Content

In 1996 Bill Gates famously stated, "Content is King," and 25 years later, he is still right. If you lack great content, it is game over, and therefore you need to improve website content continually. Great content generates strong engagement, which is why "Engagement is Queen." In this chapter, we will explore some of the basics of content.

Let us start with a true story about content improvement.

Website Redesign Gone Bad

Around Y2K, we were contacted by a prospective client that had a website disaster, and they were trying to figure out why. A designer had convinced them that they needed to **redesign their website.** The designer told them to **lose the dated folksy style** and **replace it with a world-class, state-of-the-art, cutting-edge design**. They were told they needed the latest e-commerce shopping cart with advanced analytics. They were told that they should play down the phone number because **everyone wants to buy online**. The comps used in this argument were Apple, Amazon, and many others. The client bought into the vision and agreed with the designers.

They approved a budget they could not afford, but they felt compelled to **compete at the next level.** Over the next few months, **following best practices,** the designers delivered on their promises. By all measures, this website was a huge improvement. The design was world-class and made their little business look **competitive with the biggest players** on the web. They were excited to see the launch date coming up and were planning on how to manage all the new growth that would surely follow.

But something else happened instead

The launch date came and went, and the site came online, on time, and budget. The early results showed increased traffic, but the sales dropped, and they went into panic mode. They scheduled an emergency meeting with the designers, consultants, and the sales department, but nobody could find anything wrong. Everyone they talked to say the new **site was better, slicker, and cooler**. But still, the sales were not there, so they went back to the old design, and everything returned to normal levels. They stood there in shock, trying to make sense out of what just happened.

This was when our phone rang

We listened to their story, examined the websites, and we agreed with their assessment. The new website was beautiful, slick, and well optimized. After discussing several things they had already explored, we put a crazy idea on the table. We proposed the insane idea that nobody else had thought of – **wait for it** – "Let us ask the customers." We called ten random customers and asked them what they thought of the new website.

They all agreed that the design was much better, very corporate, and made the business look much more professional. However, customers LOVED Grace and Greg and the handcrafted products and the personal service they had always gotten. One customer capsulized it perfectly **"If we wanted a low price and impersonal service, we would go to Amazon."** What customers got from Grace and Greg was a handcrafted piece of art created with love and delivered with a high level of personal service.

It turned out that the dated folksy style matched perfectly with the "Wants" of the customers. The new site looked like they had lost all of that and had gone all impersonal, professional, and corporate. A few customers even thought they had been acquired and were sorry to hear about the change. The customers started to treat their dear friends, Grace, and Greg, just like Amazon, making the sales proposition all about low price – a contest they always lost.

The Moral of the Story is

Web content quality is relative to customer's wants, expectations, and values. Content must be authentic and perfectly in sync with the rest of the customer's experience. Customers will immediately sense when a business is not being authentic, and their reaction is rarely good. Grace and Greg are fictional names, but the story is true and has happened more than a few times over the years.

Experts Learn from the Real World

At a professional level, you need to evolve and refine your skills to reach maximum engagement. Content is facts, data, and messages. Engagement is when your data morphs into **meaningful information that engages the mind of your visitor**. Engagement is difficult to achieve and even more difficult to measure.

We will make an audience assumption: you already know the basics since you are reading this book. Once you get past the foundational basics, it has been my experience that personal growth comes from networking with other professionals that challenge you. These people can be internal or external staff. As you reach the higher levels, the mix tends to shift to be more external.

Networking for Content Professionals

Content Professionals need a trusted network that will give them brutally honest feedback on their work to help them improve website content. The single best source of content improvement is professional peer feedback. A great network will force you to grow up and excel at your craft.

The natural tendency is to look internally, but that is a flawed strategy and is the first step in mutually assured failure. Internal staff rarely have agendas that align with the target consumer of your content. You must reach outside your organization and build your network. The more robust your network, the better you will be at delivering spot-on content. There are many ways for a Content Professional to expand their network, and here are some of my favorites:

Business Groups

Toastmasters focuses on live presentation skills and excels as a business networking strategy. The feedback is live, and you can learn from your efforts and the other members. I am new to this group, and I am beyond impressed. Toastmasters focuses on educating you and making you a better public speaker, but the basics apply regardless of the delivery form. The huge advantage that Toastmasters has is the live interaction, and nothing is better than being able to see the audience's reaction to your content, then using this feedback to improve your website content.

Chambers of Commerce

Almost every city has a Chamber of Commerce, and we have been an advocate of engaging with Chambers for decades. I was first introduced to Chambers and their community mission in Milwaukee in the late 80s. Just belonging creates almost no value but engaging with the Chamber is of high value. Simply put, you get out what you put in.

All Chambers have Mixer events where they bring their members together, and these are great opportunities to test your latest content and messaging. Attend a mixer to meet ten new people each meeting, and in a brief time, you will greatly expand your professional network.

Industry Trade Groups

Every business aligns with one or more trade groups. We have belonged to the regional technology trade association Softec for decades. Both our CEOs have served as President of Softec. Like all trade associations, they create connections to parts of the community that can be important to your business. If you do not currently belong to a trade group, change that today.

Leadership Networks

Within the business community, you will find specialized groups like CEO Roundtables that can be invaluable to your marketing networking because C-Level executives are a common strategic target of campaigns. These are a little tricky, and you may have to build this through others in your organization. For example, CEO Roundtables require that you be the CEO to be a member. This requirement can be a challenge if you are not a CEO. If this is the case, you need your CEO to engage with the group and provide mentoring guidance to you.

Ideal Customers or Prospects

This type of group is the toughest to build but the absolute best to have. The key to this is to make sure it is the ideal customer, not the biggest customer. Often, the biggest customers are not representative of the market and get special treatment that cannot be replicated. The challenge is how you make the relationship reciprocal and balanced, so the other person receives fair value for their time and effort. Typically, this is done by giving them an incentive like a discount or something of that nature. The ideal relationship is a reciprocal professional relationship of peers, where both value the opinion of the other.

How these Sections Connect

We started with a story about a web design disaster and then talked about building a professional network. If that customer had and used a Customer Network, they would not be part of this book. They would have caught the problem before they started the website redesign saving money and heartache.

How to improve engagement

Engagement is not a single event; it is a state of being. It flows through three steps, and your design should track these:

> - The Engaged Reader
> - The Returning Visitor
> - Hard Conversions

The Engaged Reader is a status earned when the visitor goes beyond the landing page. In our business, we count engaged readers when they leave the landing page and go to another page to learn more about our business. Google Analytics records this as a second page read within the session. We monitor that as a goal that is imported into Google Ads as a conversion.

The Returning Visitor is commonly an 80/20 rule. Check your data to see if this is true in your business.

- ➢ Do returning visitors account for less than 20%?
- ➢ Do 80% of your conversions come from returning visitors?

If your business has a relational marketing strategy, the odds are that this will be as true in your business as it is in ours. We find that 80% of our conversions come from returning visitors, not new visitors.

Conversions are when a visitor does something that the business has decided has specific business value. In a relationship business like ours and many others, these are lead forms, call requests, calls, and appointments set. We have specific targets for each of these, but they rarely happen on the first visit.

To Improve Engagement
You must become one with the Customer

If you have developed your professional network, make sure you reach out to it to evaluate your content. Ask customers how you can improve your website content. This act engages customers and provides actionable input to improve your content. Over time you will find yourself asking what a specific person in your network would say. When you can answer as they would, you are truly becoming one with your audience, and you are on your way to marketing greatness.

Top 10 Landing Page Rules

We review, evaluate, and provide design input to more landing pages than any designer we know. Our clients evaluate our services based on results, not traffic, so we take our role in landing page design very seriously. We are landing page testers, so we see what works and what does not.

Landing pages are created for prospects, not customers, and this distinction is important. A landing page is the prospect's first contact, and you only get one chance to make a first impression. That chance is your landing page. Web sites are designed to answer frequent questions and give customers additional information about your business. Landing pages are designed to answer their early questions and get the relationship advanced to the first level.

A response element is your "Call to Action" that asks for what you have decided is your objective for the landing page. There is no perfect response element, but we have developed several guidelines for evaluating landing pages.

Rule 1: Location, Location, Location

We often find the response elements buried at the end of the page like they were ashamed of asking for the order. We like to see the response element **"Above the fold"** on the right side. Above the fold is the content visible when they first land without a roll or page down. The fold is different for each monitor size, so there are many possible fold locations. It is technically possible to be below the fold in both directions. We recently saw a landing page that was so wide that the only visitors that saw the response element were those with a 22″ or larger screen. If you think you lose exposure when it is below the fold vertically, you do not even want to think about the loss when it's below the fold horizontally! A general rule of thumb is that each user interaction required on a landing page will cost you 50% of the audience. In mobile page design, the location is extra important because space is at a premium.

Rule 2: Standout

A response element should stand out on the page, and it should be clear what you want. Do NOT attempt to be clever because everyone that does not get your amazingly clever button is one less conversion. We have seen more than a few response elements on landing pages that you had to know what it was and how to respond. It might seem extremely basic, but words like "Click here to" are perfectly acceptable on a response element.

Rule 3: Offer alternative responses

A response is a response, and very few clients care how the business comes in as long as it does. Some people like to talk with people, some like email, some like online forms, and some like to visit the business. Ensure that you offer all the possible ways to contact your business. Alternative contact methods create some tracking problems, but you must ask yourself, would you rather know where a lead came from, or would you rather have the lead? Most of my clients tell me they want the lead and are willing to sacrifice some reporting ability.

Rule 4: Ask for the order early and often

There is nothing that says that you can only ask for the order once. We commonly see high-performing pages with a response element at the top of the page and another at the bottom. This rule is especially important if you are using a long copy approach to the landing page.

Rule 5: Ask only for what you must have

This one is violated all the time, and it costs you dearly in lowered responses. You do have to examine the business requirements, but many websites ask for way too much information. The insurance industry is probably the best example of this. If we call an agent and leave a message, we leave our name, phone, and a short message. If we do this on many websites, they want to complete the application information before accepting a question. Never ask for information you do not need yet, or at least make it optional.

Rule 6: Lower the commitment level

Many people create landing pages with the expectation of taking a completed order. In some cases, this is very reasonable, but ask yourself, "Do you routinely go from inquiry to order in one call?" If the answer to that is no, then your landing page should try to get to the first level in the process and not try to go from no relationship to a completed transaction in the first interaction. Another mistake made here is making a sales lead look like a committed order.

Rule 7: Talk about what the prospect values

Without rolling the page down. Are the benefits of your product or service clearly stated based on the search that the person performed? If the answer is no, then the landing page performance will suffer. One of the great advantages of Google Ads is that you know what the person asked about, and you can control where they land on your site. If we searched for Italian Shoes, clicked on your ad for Italian shoes, and then you drop me on a page unrelated to Italian Shoes, then you **do not deserve a conversion**. Very few products or services have a single sales attribute, so your landing pages should not be the same. Before you invest in dozens of landing pages for every possible type of traffic, make sure that there is enough traffic **to justify the investment**.

Rule 8: Leave your ego off the landing page

We see landing pages that use 1/3 or more of the above-the-fold space for the company logo and identification. Do you need that, or are you feeding your ego? It's a fair question because that space could be used to answer the prospect's questions.

Rule 9: Great copy sells

If you talk directly to the visitor on a subject they care about and do it engagingly; your landing page will work. A debate has raged in the direct response field for decades, and that debate is between long or short copy. We are not going to take a position on this one because we have seen both work and our recommendation is to evaluate both. Short copy generates more leads that are less qualified. Long copy generates fewer leads that are more qualified.

Rule 10: Never stop testing

You can never be done with a landing page. If there is enough traffic to justify the page development, you should never stop trying to improve the response. Different people will respond differently to different page designs, so find the best balance for your business, product, or service is a long-term process.

Rule 11: Be seamless

Landing pages are not your website, although stylistically, they should look and feel like your website because you want them to integrate seamlessly. If a visitor becomes engaged with your landing page, they will often want to know more. You do not want to duplicate all the great copy and information you developed, so the simple solution is attaching the landing page to the rest of the site using the normal site navigation.

Under Promises and Over Deliver

Did you notice that we delivered more than we promised? You got 11 ideas, while the headline of this article only promised you 10. When you deliver more than expected in the web experience, it moves your relationship with the prospect forward and starts to build the trust relationship you need. Landing pages are not as simple as these 11 rules, and these are in no way an exhaustive list or even the most important. They are simply a primer to get you thinking about this complex topic. With Google Ads, you are paying for people to visit, and you know what they are interested in, so it's almost a crime not to provide them with a great web experience so you can develop them into the customer they will become.

Search PPC Advertising

amazonadvertising

Google Search Ads Management

Google loves automation, and they have been trying to automate every facet of Google Ads since October 23, 2000, the first day of AdWords operations. They would love to have all four million advertisers just set it and forget it. Do not fall for this myth. Google Ads is and remains self-service advertising. Failing to maintain your Google Ads Account will set your credit card on fire and give your new business over to your competitors. Let's explore the basic maintenance of a Google Ads Account running in the Search Network. For those evaluating our services, what is described here is what we do for our "Google Search Ads."

In the sequence of rolling out a digital marketing plan, what we call the "Order of Battle," Google Ads is second only to a Website with phone and conversion tracking. The reason for this prioritization is that Google Ads starts fast, and you probably want fast results. It also provides data needed to compete in other areas such as Search Engine Optimization (SEO).

Google Ads does not require daily maintenance, but it is wise to glance at it each day. Doing this allows you to keep an eye on the spend, conversions, share, and anything else considered a critical KPI. These key metrics are simply taking the pulse of the account. Internally, our maintenance schedule is weekly, with most processes done monthly.

Under normal conditions, we touch each account we have under management every week but for different purposes. If you keep up with routine account maintenance, you will avoid lots of expenses.

Minimum Monthly Maintenance Items

1. Research
2. Experimentation
3. Bid Adjustments (2-4 times a month)
4. Search Term Review
5. Advertising Split Test Reviews
6. Budget and Impression Share
7. Google Analytics to Google Ads Verification
8. Quality Score Review
9. Rank Loss Review
10. Competitor Review & Changes
11. Bounce Rate Review
12. Analysis & Report

Research

Change is the one constant in Google Search. To get to an expert level in Google Ads, you must have a curious mind and consistently expose yourself to the latest writings of the best experts in the industry. A typical Google Ads Expert should expect to spend 4 hours a week researching to stay up to date. You have to monitor the Google Ads & Commerce blog and independent blogs like Search Engine Land. These are great starting points, but they are not the universe. We follow over a dozen other blogs on Ads and Search and read them weekly.

Experimentation

In Google Search Ads, you either experiment or die. Without well-crafted continual experiments, you will not learn how the system works. Without continual learning, your skills will degrade, and your career will come to a very ugly end. We like to have at least one experiment going on in every account we manage. A good Google Search Ads Expert will spend 4-6 hours each week designing, executing, and evaluating experiments in Search Ads. The funny thing about Google Search Ads is that every account is unique, but they are more alike than they are different so that you can share concepts and knowledge across accounts.

Bid Adjustments (2-4 times a month)

Google Search Ads are an auction, so the bidding is constantly changing. New competitors enter the market, existing ones launch new initiatives, and clients change strategy. All of this impacts the bidding on the account. Google has several automations in bidding, and truthfully some of them are excellent. Still, they will rarely win over an expert human adjusting for known and unknown factors in the market. We do like to take advantage of the preference given to these automation tools when they are new. Being quick to adopt a new pricing strategy that fits the business is something you want to go for.

Search Term Review

Reviewing the search terms in Google Search Ads is a key process of account maintenance. You examine each search you purchased and decide if you would buy that again. Then you engineer the change to the account to include or exclude that search. This maintenance is how you tune an account. Making those changes will not modify the past, but they can correct the future. Much of what comes out of this process are negative keywords that require great care. Throwing a negative into an account can exclude searches that you want, and they can become a silent killer of the account. In our annual Xmas letter to Saint Google, we have asked for negative search exclusions for years. Until they give that to us, you must be careful. Time spent on this task varies widely depending on account volume and complexity of the changes. This maintenance takes about 2 hours in most accounts, although we have seen these take days in some.

Advertising Split Test Reviews

We believe that all accounts with sufficient volume should always have split tests underway. With this task, you call the results of tests that have reached statistical significance and create new ads to establish the next generation of tests. Calling a test takes only a few minutes but creating the next generation can consume several hours.

Budget and Impression Share

This task evaluates the budget and any share of the market you are not yet competing for. Some clients have fixed budgets, and in those cases, this task takes a few minutes. Others control the account with the budget and a target CPA (Cost Per Action). An action is a "Measured Event of Value" the business approved. Common actions are Orders, Leads, Subscriptions, Engaged Readers, and Phone Calls. When the CPA controls an account, the evaluation here is, "Can we buy more at the same or lower cost?" When managed by CPA, it is common to take this up to make sure the CPA does not go sideways.

Google Analytics to Google Ads Verification

Google Search Ads consume cash, and every accountant on the planet will tell you that cash should be double-checked. This task verifies that the traffic that Google Ad says you bought showed up on the website. Google Analytics is a common tool for measuring the website. This verification only takes a few seconds to perform, but it is a critical checkpoint. If these numbers are off more than normal, then you need to throw a flag on the field and immediately investigate the loss. Some accountant has heart failure right now because we advised you to verify cash transactions with an error range. They are right, but you must realize that marketing data never perfectly matches because there are technical flaws in all reporting. Plus, or minus 10% is acceptable in many accounts, but that varies from account to account; it is the variation from the account's historical norm that you are looking for.

Quality Score Review

Quality Score is the other form of money in Google Search Ads, and you should treat it with the same respect you give to cash. The quality score is calculated from the relevancy of the keyword, ad copy, and landing page. It is remarkably close to the SEO score. Quality Score and SEO Score share 90% of the same attributes. If you have a quality score challenge in Google Search Ads, you have the same problem with your SEO score. Improving a quality score is a complex topic worthy of a whole book. The time this task consumes varies, so you must manage your time on this. The mystic nature of this requires a cycle of adjustment, test, and adjustment. The quality score takes time to improve, and it is not an instant gratification process. You must be smart and patient to get the job done.

Rank Loss Review

Loss of market share comes in two forms, rank or budget. Fixing the budget is easy; give the campaign more money. Loss to Rank is a complicated process. Rank is driven by the bid times the quality score. The easy one to change is the bid, but you have to look beyond the rank and look at the impact on the CPA when you do that. If the account is budget-constrained, increasing bids can **reduce** the total volume of traffic. The loss will flow from the rank over the budget, but the net results are less traffic and less business.

Competitor Review & Changes

Competitors are evil, sneaky, underhanded scoundrels that are constantly trying to take traffic that rightly belongs to your business. This task is about checking to see what the bad people are up to and reacting to what you learn. There are many signals involved in this process that must be considered. Most Google Search Ads Experts will spend 1-2 hours per month thinking about and planning to deal with competitors, although this varies based on the market's competitiveness.

Bounce Rate Review

Bounce Rate can indicate the quality and value of the purchased traffic, but it can also be misleading. Bounce rate is the percentage of visitors that landed on the page and then left the site. For many businesses, this indicates a loss of marketing investment. In most cases, the important thing to consider here is the trend of the data. The bounce rate review itself is a small task taking only a few minutes. Creating an action plan of what to do about it is where experts earn their living. Most Google Search Ads Experts will spend 1-2 hours on this task each month.

Analysis & Report

In the cycle of marketing, the two steps are:

1. What just happened?
2. What's Next?

We strongly recommend that all accounts go through an Analysis and Report cycle each month. It is important to step back and look at the month over month, year over year, same month, and year to date. Each of these trends helps the Google Search Ads Expert consider the trending of the account and relate that to the current plan. If you planned a 20% growth, but you got 5%, you need to develop a corrective action plan. As a minimum, our recommendation is a report of the CPA (Cost Per Action) and VOA (Volume of Actions) comparing actual and target with monthly and year over year trending. In summary, here is what that might look like.

Item	Target	Actual Current	Actual Prior	Y/Y
CPA	$25	$26.95	$22.25	$28.44
VOA	20	22	18	24

Fundamental analysis can and should go on forever, but you must watch out for **"Analysis Paralysis,"** where decisions never get made. Marketing is an imperfect science with variables outside of your control, so you must learn to live with that. It is common for outside managers to be confused by all the variables involved in Google Search Ads. In the example above, the reporting would need an action plan to bring the CPA back in line with the target of $25.

Top 10 Search Campaign Ideas

Search is the foundation of Google Ads, and we had written hundreds of articles on this since July 12, 2007, when we published our first blog article. There are as many ways to implement Google Search Ads as grains of sand on a beach. In this, we are going to explore our Top 10 campaign strategies.

1. Business/Industry Target
2. Radius Targeting
3. ABC Keyword
4. Single Keyword Ad Groups
5. Fishing aka Keyword Research
6. Competitor Targeting
7. Time Slicing
8. How-To Traffic
9. Device Slicing
10. Race to the Bottom

1. Business/Industry Target

The most common type of campaign is a focused keyword list of the business. In a broad sense, this divides down into Brand and Generic keywords, and the brand crushes the performance of the Generic. Branded traffic is cheaper with a great conversion rate, but there is never enough of it. The challenge with this campaign is that all your competitors know as much as you do, so finding secret keywords is unlikely. Leverage in this comes from your creativity in Quality Scores, Calls to Action, and other Customer Experience elements.

2. Concentric Circle Targeting

This strategy is strongest for local service businesses that travel to the customer's location to provide a product or service. Notable examples of this include Plumbers, HVAC, and Electricians. Using a radius from the dispatch point, the service provider reduces their travel cost and can target their bidding. In large metropolitan areas, this can be amazingly effective. We typically use three bands, 5 miles, 10 miles, and the rest of the target area. This way, we can be highly competitive in the traffic that is in their backyard.

Radius targeting can help businesses that reach well outside their local market but enjoy a home-court advantage. In our business, as an example, we provide services across the US, but we do much better in our small rural community. Because of this, we can bid and be more dominant in our local market because our close rate is much higher. Almost every business has some of this effect in their marketing.

3. ABC Keyword Lists

We find layers of keywords in every account that we refer to as A, B, C words. "A" words are the dependable, proven performers. "B" words contribute regularly but lack the volume and quality of the "A" words. "C" words are all the others. In a typical account with 100 keywords, "A" will be five words, "B" will be 10, and "C" will be 85. Budget, Bidding, and Position strategy are assigned to each section of words. "A" words always get 100% funded, "B" gets fully funded when possible, and "C" gets the remaining budget.

4. Single Keyword Ad Groups

These are just what they sound like. Starting with your A words, you create Ad Groups with one exact word match. This way, there is no compromise on the ad copy or bidding. When these are created, the keyword is placed as a negative in the other campaigns stopping cross-over traffic. With just one word in the Ad Group, your headline and body can be dedicated to perfectly matching the ad response to the keyword. Recent changes from Google on exact match logic will make this strategy less effective than it used to be.

5. Fishing aka Keyword Research

This type of campaign is designed to get new keyword ideas for an account. Using a set of broad keywords, you pick up searches to help generate new keyword ideas. As you consider each novel word, add it as a negative, so you do not get that again. As we write this, the ability to do this is degrading as Google reduces the level of data in the search query. This supplements other keyword research from competitors and open market searches. One thing that is useful here is competitor names and brands in broad or phrase with exact negatives—seeing the words surrounding competitive brands can be an eye-opener.

6. Competitor Targeting

Competitor's names and brands can be excellent quality traffic, but this strategy comes with some challenges. If the competitor has trademarks on their name or brand, they can go to Google and get those words blocked from the ad copy. Some competitors get upset when you use this strategy, so you must be ready to deal with a cease-and-desist letter or a call from an angry competitor or their lawyer. The other big challenge is that the quality score and CTR can be exceptionally low, resulting in more expensive clicks.

7. Time Slicing

In some businesses, peak periods with highly responsive traffic becomes active, and time slicing is how you leverage this. Rarely a campaign by itself; this is often an adjustment to other types of campaigns. For example, Monday morning from 8 am to 11 am is a peak period in many Home Services. This peak is because the demand for the services has been queuing up all weekend. People have arrived at their office with a honey-do list, and they often get this done during the early morning.

8. How-To Traffic

For service providers, How-to traffic can represent early-stage sales. Transitioning from how-to to a service request happens, especially when the service has a degree of risk to the work. For example, How-to install a Water heater can become a service call when the person realizes the risk of making a mistake on the install. How to traffic is not worth as much as emergency traffic, so it is less competitive. Getting your brand in front of the consumer early in the sales cycle can reduce the long-term cost of the conversion. This traffic aims to get the later searches to be your brand instead of the generic product or service. You need an exceptional website experience to convert how-to traffic into a future service request.

9. Device Targeting

Sometimes the device can change the traffic and goals of the campaign and when that is the case, then separating this makes sense. Device slicing is a bidding setting, but it can be used to separate traffic. Simply set the devices you do not want to a bid override of -100%. Devices can change the way you think about the traffic, especially when you separate desktop and mobile.

10. Race to the Bottom

Race to the bottom is one of my favorite types of campaigns, but it has requirements. There must be impression-share left, and the budget must be a constraint. Then reduce bids until the budget is no longer reached. Be careful because response rates can be related to the ad position. Closely monitor the budget and CPA. To keep bidding simple, we reduce with percentages on Devices, Locations, or Time Slices. You need to give the system a few days to adjust between bid changes before the next round of bid reductions.

Delivering More Than We Promise

We promised a Top 10 list, but we believe in delivering more than we promise, so here are a few extras.

Audience Slicing

Google allows for seeing search performance for what they refer to as In-Market Audiences (what people are actively researching) and Affinity Audiences (based on general interests). Taking this one step further, Google gives us the ability to have a Search Campaign specifically tailored to just that audience. For example, with movers, you can have a Campaign targeted to the "Moving & Relocation" In-Market audience. The goal is to reach people searching for your product or service and which Google has classified as having a general interest in the industry.

Remarketing Lists for Search Ads (RLSA)

Unless you provide emergency services, the harsh reality is that most website visitors will not contact you the first time they visit your website. With that in mind, one idea is to set up an RLSA Campaign that exclusively targets people based on which Remarketing audience(s) are set up in the Campaign targeting. While volume on this is typically modest, traffic quality is strong, and this is a terrific way to get your brand name in front of website visitors the next time they make a search relevant to your product and service.

Traffic Funneling

There are instances where you want to separate variations of a word into different ad groups or campaigns. When close variants came into play, advertisers lost the ability to direct searchers to the proper ad using keyword match types. The best way to funnel a search to the correct ad is to use negative keywords. For example, a moving company will have 'mover' and 'moving' keywords in their account, and having different ad groups for these two sets of keywords is common.

Problem: A search for 'mover' can match to 'moving' and vice versa due to close variants

You want the search to match your keyword and the keyword to match the ad. The table below shows how you would make sure this happens:

Ad Group	Negative Keyword
Mover	moving

Moving	mover

This keyword selection ensures that searches containing 'moving' do not get an ad that says mover but rather an ad that says moving (and vice versa). You want to match the keywords to the ads because the closer the search term gets to the keyword, the better your Click-through Rate. In turn, a better CTR leads to better quality scores which leads to improved Cost per Clicks.

Ad Copy as a Traffic Filter

Using ad copy to filter traffic is a surprisingly frequent problem where the client wants traffic based on keywords that are not in the search. For example, a Plumber that wants to only advertise for Water Heaters or a software developer who wants only projects of a minimum size. In both cases, the search does not have the qualifier, so there is no way to control this with keywords. The tool of choice here is to use the ad copy to qualify the traffic. Ad copy as a filter will cause your CTR and Quality Scores to suffer, but your ROI will be better. For the Plumber, the person replacing a Water Heater searches for a Plumber and City Name. Your ad headline and body need to stress Water Heaters, not general plumbing. This headline violates the foundational rule of matching the headline to the search. The headline and body need to contain the "Minimum $10k Projects for the Software Developer." Both will put the CTR and Quality Score in a free fall. To make this work, you must override the quality score with your bid, so you must be careful that the CPA you get is acceptable.

Race to the Top

Race to the Top is pricy and dangerous, but sometimes it is the right call. In this strategy, you bid to get and protect the top position. Once you are in that position, you keep increasing the bid to raise the bid against you takes a substantial number to break through. In Google Search Ads, your actual cost is $0.01 more than the bid behind you, so pushing the bid up does not cost you until a competitor responds. The real math is fancier than that, but the effect is the same. If you have a large spread between your bid and second place, they will only see more cost to them with higher bids until it runs over yours.

Wrapping Up

Your imagination and creativity only limit the number of campaign configurations. This article is our top 10, but over the years, we have tested hundreds of others. We cannot overstress the importance of experimentation, so pick a strategy and try it.

CEO's Guide to PPC

PPC is normally the first thing implemented in a Digital Marketing Plan because it is fast, measurable, and provides crucial data. The downside of PPC is the cost, and as the CEO, you must implement cost controls to make sure it is a sound investment. CEOs drive the vision and strategy and never have enough time. In this chapter, we avoid getting too deep into the weeds and stay with managing the strategy, not executing the tactics.

Recommendation Summary

For those that want just **the highlights**, here they are:

- ➢ Never take your finger off the pulse:
- ➢ Set **CPA** (Cost Per Action)
- ➢ Set **VOA** (Volume of Actions)
- ➢ Monitor Goals and Actuals Monthly
- ➢ Do NOT Set It and Forget It
- ➢ Smell the Exhaust
- ➢ Would you buy that search?

- ➢ Would your ideal customers search like this?
- ➢ Does the search include intent to engage your services or buy your products?
- ➢ Hire a PPC Coach for CEO & Account Manager

We will start today with a true story of a CEO losing control of PPC and following specific recommendations on avoiding writing a similar story in your business. All while staying at the 50,000-foot level.

Never Waste a Perfectly Good Disaster

We, unfortunately, have way too many stories where a CEO lost strategic control of PPC, creating an emergency. In this case, as it is with many, our phone rang late on a Friday afternoon. The CEO was in a panic because they were completely missing from the Google results, their phones had gone dark, and their website was offline. They had tried for two days to reach their PPC support person, but they had disappeared. While we would not learn this for several days, this account had been degrading for some time, and it was only when the website vanished that the CEO stepped in.

Emergency Declared

The website vanishing was due to the domain expiring, which was easily fixed. The company paid their bill, and the site was back up and running, but the account performance was far from healthy. We dug into their Google Ads, and their quality score was in the toilet, with the highest score being a 3. We dug deeper and found keyword-to-content misalignment. They were paying big dollars for keywords that were not even in their content. As the quality score degraded, their PPC manager simply kept increasing the bids to overpower the system. The net effect was that the client's credit card was set on fire, and their phones were dead. We paused the poorly supported keywords and rewrote ads to improve quality score relevance. By Monday, they were functional but were a long way from fully recovered.

Moving from Triage to Preventive Health

Over the next several weeks, we collaborated with the client putting keyword support into their content and bringing them back online. We deleted keywords that were attracting traffic but lacked the intent to engage their services. We reorganized keywords into narrowly defined groups and tightened up the ad content. As the quality score improved, we reduced bids gradually to bring the CPA more in line with what the Client needed to make their business model work. We discovered more deferred maintenance, including several website names purchased as part of an SEO strategy. These site names had different registrars, hosts, and redirects. This nightmare took a few days to get our hands around. The SEO manager had implemented several tricks that Google has long since figured out how to detect. We suspected that this was further impacting current organic traffic. A substandard quality score is a good indicator of SEO score problems, so we unraveled these tricks as well. Within 30 days, SEO traffic started to return, and odds were good that Google slowly but surely forgave them for past sins.

This story and dozens more, just like it, has a common theme. The damage happened over a long time in silent killer mode. These conditions might have continued except for the blessing in disguise of the domain expiring. If that had not happened, this walking dead website might still be causing damage and burning cash. Below are some simple steps CEOs can implement today that will stop you from owning your very own Google Ads Disaster story.

Never take your finger off the pulse of PPC

Sales is a game of numbers, and in PPC, the big ones are CPA (Cost Per Action) and VOA (Volume of Actions). These should be on your monthly CEO Dashboard. Actions are things like Orders, Customer Acquisitions, Conversions, Leads, and Subscriptions. Anything you deem of value should be measured as an action. Do not let acronyms or industry jargon confuse you; they are all the same thing.

Running a Google Ads Account without a target CPA and VOA is a disaster looking for a place to happen. The commonsense rule is **"No Target CPA & VOA – No Money,"** but as we all know, common sense is not always that common. Check the actual goals each month and **insist on a correction plan** if they are headed in the wrong direction.

Some CEOs think that CPA is a form of expense, and like all expenses, it should be driven from the P&L by the entire management team, using pitchforks and torches. The reality is that CPA is an investment tied to the value of the action. Target CPAs must be **as high as necessary but no higher than that.** Target CPAs exist on the spectrum of cost control and revenue. It is equally as bad to push the CPA too low or too high. It needs to be just right.

A low CPA risks loss of revenue, while a high CPA causes declining profits or even bankruptcy. CPA falls into the **Goldilocks principle – not too high, not too low, just right.** Many things can move the CPA, but let's stay out of those weeds. Make sure you have a target CPA and that your staff reports on that regularly. If it goes out of balance, insist on a correction plan to bring it back. If you need help getting to your target CPA, give us a call. We would be glad to assist in this process.

VOA (**V**olume **O**f **A**ctions) is closely related to the CPA. Every business plan has, at its core, a revenue target, and for that, you need a certain amount of new business. Any CEO worthy of the title will know their target. Most initial actions are often one or more steps removed from direct revenue, so the VOA often must be calculated.

For example, if you need 40 new clients each month to meet your revenue targets and your Sales Department closes 50% of their leads. That means you need 80 new leads. With your CPA and VOA in hand, the budget is simple math:

CPA x VOA = Budget
Do NOT set it and forget it

CEOs get lured into thinking that PPC is a project with a start and stop date without the need for maintenance. This myth is put forward by software developers who are not Google Ads Experts. These developers have never had to turn a profit in PPC, and they think their software is way smarter than it is. Artificial Intelligence, Machine Learning, and all their software cousins are a long way from understanding the nuances of language and human interaction in your market.

PPC is a never-ending and evolving process that is an integral part of the business ecosystem. PPC exists at the junction of people, language, and systems, and it isn't easy. If you set it and forget it, it will eventually lose its mind and **drive full speed into the closest immovable object**. Unfortunately, many people, including Google, will attempt to convince CEOs that their software cures this – **they are lying to you**. Set it and forget it is a strategy that creates an **accident looking for a place to happen.**

PPC is more than just Google Ads, but we will use that as the example network. Each year, Google announces hundreds of changes, and nobody knows how many are unannounced, but we know it is a lot. Each change represents an opportunity or risk or both to your account. If Google does not change the rules, your competitors certainly will. If you have a strong, profitable position, your competitor will not roll over and let you have it. They will fight, and no account can stand there and take blow after blow without experiencing pain.

PPC is complicated, but as a CEO, you must make sure you hire smart people and guide them on the vision and goals. People are people, and you must hold them to higher performance than they would hold themselves. This accountability is the role of the CEO.

Smell the Exhaust

Auto mechanics use an analyzer that sniffs the exhaust and provides engine performance insights. In Google Ads, we have "Search Terms," which provides insights into performance. Require your PPC person to send a report of your top 100 search terms every month.

Gathering these data takes only seconds to create and export to a spreadsheet that they can email to you. Then spend five minutes looking at the searches and ask yourself:

1. Would you buy that search?
2. Would your ideal customers search like this?
3. Does the search include intent to engage your services or buy your products?

Then ask your Account Manager about any that get a 'no' from the questions above. Do that every month, and your account will become more tuned to your vision. You will quickly learn that you only need to look at the Top 10 and ask questions to keep your PPC direction aligned with the business vision and goals. Many CEOs find that they can get this down to just a few minutes each month.

Hire a PPC Coach

The first question is **"How Important is PPC?"** If the answer is some form of high or extremely high, then consider hiring a PPC Coach. How much you engage your coach depends on how far you want to go and how fast you want to get there. As CEO, you oversee the person that has day-to-day account management responsibilities, either internal or external. You do not need to be smarter than they are, but you must understand enough to appreciate the value they bring to the business. You also need to be up to date on the PPC opportunities and risks. Most CEOs can get this done within an **hour a month,** but they often start with an **hour a week**. You can start weekly and then back it down to once a month to stay up to date. Some companies keep the coach once a week, but they move 75% of those hours to the account manager, reserving one hour for an executive update briefing. This tactic expands the resources of the employee managing the account and gives them a professional second opinion. A side benefit of this is a trusted outside PPC auditor and backup staffing if the employee resigns or fails to meet their sales goals.

Vision & Strategy

CEOs direct and control the vision and strategy of PPC, but they are rarely PPC experts. By monitoring a couple of key points and asking tough questions, you can control the strategy and delegate the tactics. If the CEO spends more than a few hours each month in PPC, they are too deep into the weeds. Control the strategy but do not get sucked into the details because they never end, and you have a company to run.

Emergency Traffic

Emergency Traffic is searches related to a problem that is actively

leaking or spewing on the floor. The person doing this search will continue to shop until they are confident that someone will arrive quickly and solve their problem. This section uses Plumbers as the emergency service example, but the same is true in any emergency service.

Rule 1: With the first glance at your mobile site, it should be clear that you are the emergency service they seek.

Emergency traffic is expensive because it is highly responsive, and everyone knows this. Competitors will not roll over, and battles over this traffic are epic. Treat emergency traffic with respect, or it will set your credit card on fire.

Get Simple and Be Direct

Emergency traffic is event-driven, meaning that your ad serves the demand; it does not create it. There is no 'make a pipe burst' button in Google Ads, and there never will be. To serve this person, you want to be as clear and direct as possible and avoid any temptation to be clever or humorous. Be as clear as you can be that you can serve them and serve them now. Headlines like Plumber 24/7 and the "City Name" are common.

This headline is a quick and simple read. Then spend some time thinking about how you would talk to this person and put that in the body of the ad.

Position Matters

In emergency traffic, position matters, and it matters a lot! For this type of traffic, 2nd place is better known as the first loser. On February 26, 2019, Google announced that they would be sunsetting Average Position in the Google Ads System. Translated to English, this means they are doing away with providing the data point of your Average Position and replacing it with Absolute Top and Top Position Percentages. Most of what you used to know about the position is no longer true. Google claims that:

"These new metrics give you a much clearer view of your prominence on the page than average position does." There is a degree of truth to this,

but it also changes how we must talk about this data. Being number one is no longer the goal. The goal is to be in the position that best serves the business because being first can often be too expensive. The position is a balance between performance and cost with a VOA consideration. As you increase your bid, the position will improve, but the cost per click will go up.

The **consumer is under stress,** and they will stop shopping as soon as they reach a business they believe will solve their problem. Note that we said "believe" because this comes from the initial contact with the consumer. If your phone goes to voicemail, the consumer will go on to your competitor. If the person answering the phone cannot satisfy them that their problem will be fixed, they will move on to your competitor. If you have an online scheduling system and the system fails to confirm the appointment within seconds, the consumer will go on to the next competitor. It is expensive to be first in the paid results, so please do not waste that money and blame Google Ads when you fail to close the deal. We have seen a lack of training for the in-bound phone staff, which cost businesses thousands of dollars a month.

Get Local

Emergency services are all about being local, and the more you can make that point, the better. With recent changes in the Google Ads location process, it is now possible to localize beyond the city by indicating a Service Area. In many parts of the country, there are regions known by and spoken about by locals different from the city name. For example, in our area, locals do not say "San Luis Obispo"; they say "SLO" pronounced slow. This local dialog creates some funny expressions. Slow plumbing is what you call a plumber to fix but not here. There are also region names, and locally we have a South County, North County, North Coast, and Five Cities. If you look at a map or use a city list, you will miss these keyword opportunities. The secret is to think like a local.

Think Mobile be Local

When it comes to emergency services, mobile has taken over the world. Desktop is still important, but you must **think mobile-first** and get hyper-local. In the olden days, businesses benefited from a toll-free number, but today, an area code and phone prefix scream local. The 805-

area code covers a massive territory, so it says local but in an unbelievably soft voice. If you want to scream local, get phone numbers with the right prefix or office code. Locally, an office code of 481 or 489 is the Five Cities aka South County, and 541 is San Luis Obispo (City). San Luis Obispo is the name of a city and a county. Each of these areas is considered separate from the other when it comes to emergency services. An ad with a central phone office of 541 is quite different from 481, and the response will be different. Understanding the local office codes takes a little bit of research, but it is not rocket science.

Emergency Keywords

Emergency Keywords includes both specific and implied emergency. Examples of specific emergency keywords are:

- ➤ Your City + Emergency Plumber
- ➤ Your City + 24-hour Plumber
- ➤ Your City + 24/7 Plumber

Here are a few examples of the implied:

- ➤ Your City + plumber
- ➤ Your City + plumbing service

> ➤ Your City + plumbing contractor

The Your City variable should always include "Near Me" and other local qualifiers. Emergency keywords are expensive because they convert to phone calls, and everyone knows that. For our example, let us look at "Plumbing." When used with a qualifier like service or contractor, it is a great search. Without that, it is what we call a gray word. Plumbing has more off-topic traffic because of its use in DIY and Parts.

Voice is to Mobile as Air is to Life

Smart Phones have taken over the world of emergency plumbing calls. Keyword searches are morphing into sentences driven by voice search and voice assistants like Google Home and Amazon Echo. In your search data, your voice searches stand out because of their length. If the search is over six words, it is probably voice. Google has been limiting the search queries' details for years, and many qualifiers no longer show in the details.

Doing the Right Job

1. Your ad's job is to get them to your website
2. Your Website's job is to get them to call
3. Your Phone must bring home the business

It is scary how many times poor phone skills destroy the excellent work of steps 1 & 2. The person answering the phone is often the lowest ranking and least skilled person in the office and often with little or no training. Think of the strategy here! You take your largest opportunity, and you give it to your weakest player. It is expensive to make the phone ring. Please do not lose that investment because of poor training. Teach and coach them on how to:

1. Professionally answer the phone
2. Qualify the opportunity
3. Dispatch the right person to close the deal
4. Answer Questions
5. Sense Opportunity
6. Project confidence that problem will be solved
7. Communicate Dispatch
8. Follow up after the service call

There are lots of basics on how to answer a business call. They should communicate the business name, their name, and an offer to help. Inbound Phone Sales is a profession and should be treated as such. Money invested in the person's sales skills answering your phone will come back to you many times over.

Golden Rule: Your customer service is what closes the deal

Get Your Timing Right

Emergency traffic is expensive, and you need to make sure that you respond quickly and with authority. Within Google Ads, you can put timing controls in place and even modify the bids by time.

Why Can't I Run Google Ads Myself?

The short answer is you can, but the real question is, should you? Ask yourself, is your business better served by staying focused on the business of your business or Google Ads? Are you going to dedicate the time necessary to learn Google Ads? Google has invested millions trying to make it simple enough for anyone to run. However, like any profession, the devil is in the details. It seems easy enough - you create a campaign, write an ad, select a few keywords, and run Google Ads. In this basic set up Google is getting their money. Unfortunately, it is not creating maximum value for your business.

I have made my living as a search engine marketing expert since 1994. In all that time, we have never had a client who did not know their primary keywords and more about their business than we do. However, they rarely understood the complex issues involved in effectively targeting those keywords.

Plumber keywords are a classic example because the typical plumber should **only target searches with** "Intent to Engage Services." After all, many searches will be either non-specific or off-topic. The keyword **Plumber Los Angeles** is a good example. At first glance, this would seem to be a great keyword, but what if the real search is **Plumber Jobs in Los Angeles**? This search matches the **Plumber Los Angeles** keyword, and most plumbers are paying big money for that search and many others of questionable value. Google Ads seems simple enough until you get into the details. **There are three major networks:**

1. Google Search
2. Search Partners
3. Display Network

Six types of keywords:

1. Positive Broad
2. Negative Broad
3. Positive Phrase
4. Negative Phrase
5. Positive Exact
6. Negative Exact

Matching is not just to keywords because you also have keyword themes, direct placement, and remarketing audiences in the Display Network. More on this in the Display section. The keywords are a selection model for a system with billions of searches per day, and the system is an interactive auction with literally thousands of variables involved. As a person trained as a computer programmer, I can tell you without hesitation that Google Ads is the most complex system you are likely to ever run into. You never know what a person with a blank search box will type in.

Marketing is an ecosystem, and Google Ads has a role that interacts with other parts of the process. Google Ads creates traffic to your website, but the website must bring the person to action and cause them to contact your business. That is where you get the value for your investment.

The question is not, **"Can you run Google Ads?"** Because you can. The question is, **"Should you?"** For many businesses, staff will create more value by focusing on the business of the business. This value conversion is especially true in high-value services like Lawyers, Plumbers, and many others.

Keywords

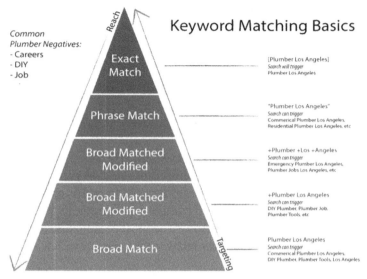

Common
Plumber Negatives:
- Careers
- DIY
- Job

Keyword Matching Basics

Pyramid Level	Example
Exact Match	[Plumber Los Angeles] *Search will trigger* Plumber Los Angeles
Phrase Match	"Plumber Los Angeles" *Search can trigger* Commerical Plumber Los Angeles, Residential Plumber Los Angeles, etc
Broad Matched Modified	+Plumber +Los +Angeles *Search can trigger* Emergency Plumber Los Angeles, Plumber Jobs Los Angeles, etc
Broad Matched Modified	+Plumber Los Angeles *Search can trigger* DIY Plumber, Plumber Job, Plumber Tools, etc
Broad Match	Plumber Los Angeles *Search can trigger* Commerical Plumber Los Angeles, DIY Plumber, Plumber Tools, Los Angeles

Keywords can be thought of as a programming language used by Google Ads to decide if the ad is eligible to show. Once it is eligible, the bid and quality score determine the Ad Rank, which is the order the ads will appear in. Matching is a series of positive and negative keywords. People often struggle with the word types and negatives used to get just the searches they want.

We will use a Plumber as the example for discussion here, but every business will have the same conceptual issues to think about. Plumber keywords exist in a universe that includes the service intent searches that a **plumber wants**. But also, mixed in these ideal searches are products, how-to/DIY, job/career searches, and others that a **smart plumber does not want**. The following pages are examples of searches that match "Plumber Los Angeles" some are great buys, and others are a complete waste of money. If you have the keyword without the 263+ negative keywords properly typed, much of the traffic you will get is not what you think. The 263 number is not just made up; it is the starting number of negative keywords for any plumber using Google Ads before adding localized negatives. The ideal keyword model will focus on the service searches while avoiding the other categories. Since the match would be more relevant, it will drive up the click-through rate improving the quality score resulting in less expensive keywords

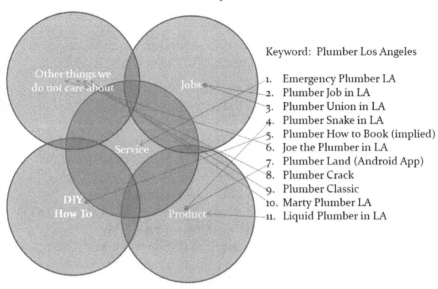

Plumber's Keyword Universe

Keyword: Plumber Los Angeles
1. Emergency Plumber LA
2. Plumber Job in LA
3. Plumber Union in LA
4. Plumber Snake in LA
5. Plumber How to Book (implied)
6. Joe the Plumber in LA
7. Plumber Land (Android App)
8. Plumber Crack
9. Plumber Classic
10. Marty Plumber LA
11. Liquid Plumber in LA

Let us walk through this simple example

1. **We Want THIS one**.
2. Jobseeker – employment traffic cost 95% less than plumbing traffic. If you want to recruit employees, there are cheaper ways to do that.
3. Another job-related search
4. A product search is better suited to Lowes or Home Depot.
5. DIY search with possible secondary value if they need a professional to fix or finish what they started.
6. Slang name made famous in a Presidential Campaign
7. A phone app name
8. Class of joke
9. Event Name or Industry Term
10. Person's Name
11. Product Name

Some of these searches are low-volume traffic, but Plumber keyword costs can become real money real fast.

Understanding and selecting the right keywords is the secret to success in Google Ads. This process only seems simple until you get started. Keywords exist in a universe where they change meaning based on their use, and search engines make this even worse by removing much of the context. Most searchers do not pose a complete question but rather enter the essence of their intent and are expressed as a base subject and qualifier. Here are the Top 10 qualifiers for plumbers:

1. **Geographic Qualifier:** Typically a city or regional name, and they are common service searches such as "Plumbers in Los Angeles."
2. **Information seekers:** These are words like **Reviews** or many others that indicate the person is seeking information and are likely to be earlier in the purchase cycle. Some plumbers like this traffic, others do not.
3. **Action Indicators:** Words like **Price** or **Quote** indicates that the person is later in the purchase cycle and is probably in the service provider selection stage of the purchase. Price will sometimes indicate a product search, but that is not always the case.
4. **Emergency:** Words like **Emergency** or **24 hours** indicate the person is looking for immediate service. These keywords often sell at a premium in Google Ads, and they tend to be extremely low volume with great intent to contract services.
5. **DIY (Do it yourself):** Includes words like **instructions, manual,** or **how-to** and indicates that the person is seeking instruction more than service. These are often high volume and low value to a Plumbing business but some plumbers like this traffic because they often get jobs when the DIY project goes badly.
6. **Products:** This one is tricky in plumbing because its intent value varies based on the product involved. Some products have low service values, such as faucets, while others like water heaters are more likely installed by a professional.
7. **Service Indicator:** Words like **Licensed** increase the intent toward a service provider, making them high value and low volume. Examples would be **Licensed Plumber,** although

with this term you must be careful of jobs traffic. A better example would be **"Licensed Plumber for."**

8. **Education:** Words like **licensing**, **course**, **class**, **apprentice**, **certification**, and others that change the meaning of the base subject and shift the intent of the keywords towards schools and away from plumbers.

9. **Careers/Jobs:** Words like **help-wanted**, **jobs**, and others that indicate the searcher is seeking employment in the plumbing field can be high volume and low quality for a plumber.

10. **Branded Traffic:** Driven by your brand or the brands of your product partners, this traffic is often highly responsive, especially if the brand is hard to find locally or if you have exclusive territory rights to sell and install.

These 10 points are just the tip of the iceberg, and how they interact with each other can be even more fun. For a Plumber, our goal would be toward the service indicator and away from the others. Other businesses might be the reverse of that.

Finding the right balance can be more art than science because language and intent are difficult concepts. We commonly find that 5% of keywords create highly profitable and predictable leads but not enough volume to support the business. To fill the volume, you must go into the less qualified words. Time, testing, and experience will help you find the right mix to keep your business running right.

Plumber Keyword Model

We are using Plumbers as the example in this book. The industry drives the actual words but not the model. The same concepts apply to your industry keywords.

The Plumber keyword model includes the generic terms that are the high-volume words. Starting with Plumber + City Name keywords are extended by specialized keywords. The volume is split 80/20, with 80% of the traffic coming from generic terms and 20% from the specialized categories. Not every plumber performs all these services, so the first step is to identify all the services that are important enough to invest in.

We start with these rank definitions:

Rank	Definition	Remarks
0	No Value	Not worth advertising
1	Low Value	Limit bids
2	Average Value	Target secondary location on the SERP and keep impression share at 50% or less.
3	High Value	Target Top Position and high impression share
4	Very High Value	Position 1 at all costs – be careful about putting a word in this class.

Then rank the Areas of Services ------------------------------->>>>

Area of Services	Install	Repair	Emergency
Backflow Testing			
Bathroom Remodels & Renovations			
Bathtub			███
Boilers			
Burst Pipe Repair			
Dishwasher			
Drain Cleaning & Clogs			
Earthquake Valves			
Faucets			
Fire Sprinklers			
French Drains			
Garbage Disposals			
Gas Leak Detection			
Gas Leak Repair			
Gas Lines			
Grease Trap Cleaning			
Grease Traps			
Green (Eco-friendly) Plumbing			
Hot Water Recirculating			
Hydro-jetting			
Ice Maker lines			
Kitchen Remodels & Renovations		███	
Laundry			
Pipe Fitting			
Pipe Lining			
Pipe Thawing			
Pipe: Cast Iron			
Pipe: Copper			
Pipe: Plastic			
Pipe: Rubber			
Pipe: Steel			
Plumber - City			
Plumber - Generic			
Plumber- Emergency			

The rank gives us the bidding priority. There are business insights you can gain from the exercise. Rank can help reserve your budget for the high-value traffic and spend the balance on lower-ranked services.

The blacked-out boxes are illogical combinations on the following chart, and all the open boxes should have a value from the table below. Zero-ranked items are a good source of negative keywords to help tune that side of the bidding. Different industries will have different service areas, but conceptually this works for most industries.

The reality in plumbing is that the word plumber and the city name is 80% of the total traffic, and everyone knows that. This 80/20 rule exists in most industries, and it drives keyword bids sky high. The advantage in Google Ads often comes from lower volume lesser-known searches.

More About Keywords

Words exist in a range of intent that we refer to as **word classes**. In plumbing, for example, we have at least; service, job/career, product, DIY, and other industries. When you examine a word, you must think about how much of the word exists in each class. For example, Water Heater will tend toward service because it is less likely to be owner-installed than a Faucet. By itself, a Water Heater has both service and product values, so we need to look at how it is extended beyond the base keyword. If you extend "Water Heater" with price keywords, that moves the intention towards being a product. If you extend it with an Installation, the keyword moves towards service.

Base Broad Searches: These are the easy ones, but they are often too expensive, and they contain too much contamination from off-topic areas. An example of this would be buying the word Plumber. Sure, it is on topic, but it is just in too many searches that you do not want.

Service Search: This is classically what the home services businesses are after, but the cost can get out of control. You must make sure that you only buy the searches you want. Most of the time, you are evaluating how much service content is in the search being performed.

DIY – Do It Yourself: DIY traffic rarely converts for a service provider, but it can and does happen. Some of the extensions that get you to DIY are things like **how-to, manual, instructions,** and other keywords like that. For the few that do convert for service providers, they are usually technically difficult products. Conversion from DIY to Service happens when you provide the how to do it information, and they will decide it is better to pay a professional. DIY to Service is unlikely with a small fix but frequently happens when you educate them on the risks.

Product Searches: Product-based searches can and do convert to services, but you must watch for extenders that indicate they are looking to buy the product. Things like "price" push toward the product, but "quotes" and "estimates" push more toward services. A base product word like "Water Heater" used with service words like "installation" is a home run in the parts area.

Job/Career: This is a class of words that tears right into the heart of some great service searches, and the issue comes from the difference between a keyword and a search query. The keyword is what you specified to match to. The search query is what the person entered into the search engine. The Classic example is **Plumber Job Los Angeles** which matches the extremely popular **Plumber Los Angeles** search. Not putting "job" into the negatives for this search can get you some expensive resumes.

Other Industries: This is one that advertisers often overlook, and it is both a risk and an opportunity. On the risk side, you can have keywords where the other industry dominates the word, so your keyword starts weak. An example of this would be words like **fraud** or **claims**. Examples would be **Plumber Damage Claims**, which will get the Plumber into competition with Personal Injury Lawyer and Home Restoration Business. The problem is that these businesses outbid you. The opportunity example is remodeling. While this is another industry, an advertisement that encourages the use of a licensed plumber can produce satisfactory results.

Writing Ad Copy

Seattle Metro Plumber - Call Us Now For Fast Response
Ad www.example.com
Let us be your go-to plumber when the urgent need arises

The purpose of the ad copy is not to sell your product or service to the consumer.

Like most things in life, the devil is in the details, and there is no shortage of details in Google Ads. The purpose of the ad copy is not to sell your business to the consumer. Let me say that one more time.

The purpose of the ad is to get a qualified consumer to click on your ad in the right frame of mind so your website can continue the conversation. The ad copy is of utmost importance because you never get a second chance to make a first impression. In most cases, the best way to do this is to take the clue from the search and use the ad to write a response to their query. You must think about this in a conversation form, with the ad being the response and the website being the continuation of that conversation. The goal is to educate the consumer on the value that your business brings to them to address the problem they started with.

Ad copy is an area that has recently seen lots of change as Google dropped the right-side ads and changed the technical limits. They are changing the basis of design to be mobile-first rather than the desktop-first design of the past. The newer format text Ad has two headlines of

30 and a body of 90. Our recommendation is to put the keyword in the headline, with a call to action in the second headline, followed by more call-to-action words. It is essential that you get to the point. You do not need to be cute, clever, or overly creative as your ads are being presented to people who have already expressed an interest based on the keyword match.

As you begin to write your ad copy, it is best to give the people what they want! In many cases, if someone is going to gloss over any part of the ad, it is likely to be the 90-character description section. Make sure that your two 30-character headlines address the searcher's need. The ad structure that we have found effective is to use the first headline to establish what you offer/locality (i.e., Plumber + City Name), and the second headline to highlight a call to action such as "Call for a Free Quote." Then use the 80-character description to support the points from the two headlines.

In life, sometimes boring is good, and this principle applies in Google Ads. The more boring and straightforward your ad is, the better it will do. If you are a Plumber in Seattle and someone searches for "Plumber Seattle," your ad better reference being a Plumber in Seattle. If you decide to be creative and go off-topic, this will likely negatively impact your CTR, Quality Score, and Cost Per Click.

Unless the search is for your business name, you do not want your company name in the headline. If the search was for a generic word, do not lead with your company name. Serving them an ad that's focused on your company name and not on your services is not helpful to either the searcher or yourself. The searcher doesn't get what they searched for, and you don't get a visitor to your website. In addition, the searcher still doesn't even know what you do because you decided to lead with your name as opposed to how you can help them solve their problem.

The "secret" of getting your ads to work in Google Ads is to get to the point. Use the keyword as a clue to their interests and address that interest. When the keywords are tightly clustered and both headlines are on-target, the results can be amazing, even if the headline is boring.

Split Testing

Split testing is an essential skill for a Google Ads Expert professionally managing an account in today's world. This chapter will explore some of the details that need to be considered to run a professional split test. We could write an entire book on this topic and still not cover it all, so for this chapter, we will limit our scope to split tests conducted in the search network of Google Ads using text ads.

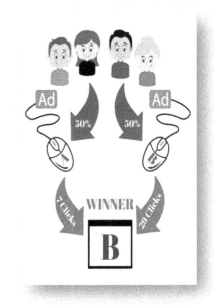

The first thing to consider is the goal of the split test; they can be designed to:

> ➢ Learn something new
> ➢ Experiment with a breakthrough idea
> ➢ Evaluate the Margin of Error

Why do we bother with split tests? Splits help us evaluate what we think is true against what is happening in the specific market. With over 20 years in this field, we cannot tell you the number of times that market response has surprised a wide range of experts. Split test results are clues, not facts. It takes lots of tests to morph a clue into a fact because there are many factors to consider.

A/B Tests – Learning Something

Technically, all split tests that run two ads are an A/B test. For this chapter, an A/B test is evaluating a single variable. Ads evolve in small steps because you can only change one thing at a time. To reiterate, CHANGE ONLY ONE THING per test. Using an A/B strategy requires a long-term view and the will to stay the course. If you are part of the instant gratification society, it is unlikely you will be successful with a true A/B approach.

Multi-variant – swinging for the fences!

When most people talk about split tests, this is the type of test they mean. They write completely new ad copy and put it in competition with the current winner. The challenge is that when the new ad wins, there are so many differences that you must guess what caused the result. If what you need is a breakthrough ad, this is the type of test to run, but you must accept that this is equivalent to swinging for the fences. If it works, it is a home run, but it is risky and is often a strikeout. When you are trying to find the message that resonates with your audience, this is a valid strategy. As the account matures, this makes less sense.

Cross Traffic Tests

Sometimes an account simply lacks the volume to do a test in a reasonable period but has enough traffic at an account level. In these cases, you can sometimes engineer tests that tie all or most of the traffic together. The key to this is creating ad copy that is generic enough to run the same headline or body in all the groups. The compromise here is that the more generic the copy, the weaker the ad tends to be, so getting the test at all costs can sometimes be too much. While the headline is giving up too much, the body text can be evaluated.

Margin of Error

A margin of error test is what it sounds like. Everything in split testing is a clue, not a fact, and you have to understand the range of accuracy you are dealing with. In a margin of error test, you run two identical ads and run them against each other. This test reveals how much the data might be off, based on the variables you **do not** control. When this is done several times, a range of errors will become visible, and you can use that to decide how much you trust the result. Over the years, we have noticed that this is about plus or minus 10% of the tested rate. An account with a 2% CTR will come in from 1.8% to 2.2% simply from the margin of error. If the result of a test is within this range, **then it is too close to call.**

Common Mistake 1 – Confusing clues with facts

Clues are data that indicate support for a theory but are not proof. A fact proves that a theory is universally true. In marketing, clues are common, and facts are exceptionally rare. How rare, you ask? Good question: In the last two decades in digital marketing, we have yet to find a fact, but we are still looking. The more clues that support your theory, the stronger your belief should become. Reaching fact-level confidence is a very long road.

Common Mistake 2 – **Not considering outside events**

Google Ads does not operate in a vacuum. The best example we can give to you is a snow removal business that wanted more sales leads. No matter how much they advertised and promoted the business, we just could not move the needle. Then a huge snowstorm came, and the phones lit up like a Christmas tree. The truth is that search marketing serves demand but rarely creates it. Remember this simple fact. The consumer "want" created the search query. Without the want, there is no search.

Common Mistake 3 – **Thinking body text counts**

In most cases, the body text (two lines under the headline) simply does not change the results. We think the reason is that people search, see the headline they need, stop reading, and click on the ad. We have evaluated this many times, and changes in the headline create meaningful change in the traffic, but the body text does not. We often see people struggling for hours to create the best copy possible, often for little or no return for their efforts. Invest your time in testing headlines and write the informational copy for the minority of the audience that read those lines.

Common Mistake 4 – Not letting the data build

Many people lack the patience to conduct testing. They want to run testing over the weekend and make big decisions on Monday morning. The reality is that most tests should run for at least a month. If you are testing a CTR with a normal range of about two percent, you will need more than 400 clicks to get valid data. There are many A/B testing calculators on the web, and here is one:

https://neilpatel.com/ab-testing-calculator/

Stealing business from Competitors

Not everyone will be comfortable with this topic because they have ethical concerns. We are at our core sales-driven, and we consider competitors to be the enemy. We have no reservations about targeting competitors' traffic.

It has been my experience that competitors, especially successful ones, are smart. We respect them, but it is our job to beat them. If they slip, we want to be right there to pick up the advantage. This strategy has positive and negative considerations, so think carefully before going after their business.

Like all businesses, your competitors enjoy some degree of branded traffic, where the search is specifically for their business name or brand. You can advertise for this traffic, so your ad appears when someone searches for the competitor.

The keywords are simply the names and brands of your competitors. In most cases, you want to express this as a phrase search so that the keywords will be in quotes like this "Rocket Plumbing." The phrase search will match anything that contains the phrase. Since you are not Rocket Plumbing, your quality score will suffer, one of the problems with this strategy. You can expect in most cases to see lots of 3s or even lower. A competitive position for this search will require a high bid because of the low-quality score.

Your ad copy needs to take a different approach because of the mindset of the searcher. If they are searching for a brand, then you need to pitch an alternative. Offer a second opinion, and this one is especially true if the search includes the word "review." Just because you can bid for the word, that does not mean you can use the work in the ad. If the competitor's brand is trademarked, as many are, you cannot use it in the ad copy. Google will disapprove of an ad that uses a trademark without permission. If you have a trademark and want to protect it, you can go through a registration process to have Google block use within the visible ad copy.

Your click-through rate will suffer because you are off-topic for most of the searches. However, there is information that can be valuable to your planning. Look at the impression levels, and you see how popular your competitor brands are.

You need to overbid this traffic to compensate for the low-quality score, driven by the low click-through rate. If you bid low, the ads will never go active.

This strategy is not one that we are a major fan of, but we have implemented this many times over the years. The downside is that the traffic is expensive, and the click volume is low.

Target Position 2

Within ad groups that target competitive names and brands, you can use this as an offensive weapon. Some competitors will guard their brands with high bids, and you can use this to drive their costs up. Simply bid up just short of beating their bid, and it will cost them one penny more than your bid. There is danger in this if the competitor is smart, they can suddenly drop their bid pushing your costs up. We have never found this to be especially useful, but I am asked about it all the time.

Budget Constrained Campaigns

Budget-constrained Search campaigns are one of the most common problems in Google Ads. Many signals are indicating you have this issue. Google thinks this is so important that you cannot even remove the notice. Google wants you to fix a low budget and constantly reminds you to increase the budget.

We think there is a smarter way to manage this problem, and it certainly does not align with the Google solution.

Budget	Status
$30.00/...	Limited by budget 〰️

Assess the Opportunity

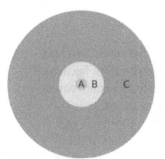

Almost every account has the same keyword mix challenge. We like to divide keywords into classes, and here are the definitions:

➤ A = Top 5% of words
➤ B = Next 15% of words
➤ C = The bottom 80%

The problem with "A" words is that there is never enough volume, and they are costly because your competitors likely use them as well. "B" Words are a much higher volume with lower quality but still fierce competition. "C" Words are the rest and typically create value at lower levels than A & B. The chart to the left is a graphic of what this looks like.

How Google Wants You to Solve This

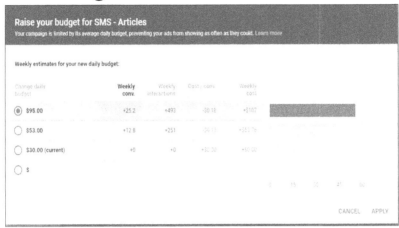

The Google solution to this problem is that you should spend more money. Let us just say that our goals align with yours and are quite different from the Google goal. We want to spend less and get more when the CPA (Cost Per Acquisition) goal is reached. If that aligns with your goals, then we wrote this chapter for you.

Keyword Parts & Pieces

"A" Words always follow the format with a subject followed by intent followed by a qualifier. Here is what that looks like:

Google Ads Expert San Luis Obispo

"Google Ads" is the subject, "Expert" is the qualifier, and "San Luis Obispo" is the location. If we get this search, it is a grand slam because it is what we do and what the person is seeking, not only an expert but a local expert. Since we are the only Google Ads Agency in San Luis Obispo County, this is a trifecta for our business. The problem is that San Luis Obispo is a small rural community, and the number of these searches is approaching zero. The person doing this search is probably someone we have had coffee with.

Strip away the geographic qualifier, and then we have "Google Ads Expert." A higher volume keyword, but the competition just increased tenfold. We have no dumb competitors in this space, so we can tell you without question that every competitor knows this is a word worth going to War over.

Strip away "Expert," and you get the base subject, which has massive volume with questionable quality, intent, response, and value. If you try to squeeze the last drop of traffic, then these are worth going after—if your campaign is budget-constrained, then putting this keyword into your campaign a bad move.

Quality versus Quantity

Like most intricate system problems, you face balance issues. The answers are not right or wrong but a balance between quality and quantity. If you tighten the quality, the quantity will drop, and with a budget constraint, the goal is to find the balance. You want your campaign resources (money) to go to the highest-ranked keywords.

The Mechanics of this Change

Several variables are going on as you move through this strategy, and it is like peeling an onion. As you move through each layer, the data will expose itself. For every layer, you will have a new set of challenges. Initially, we start with a big bucket of keywords separated by theme. As the data evolves, keywords move into their volume classes, and the ad copy gets even more specific to the keyword. This evolution of the account takes time depending on the volume of the data passing through the account. The more data you have, the faster you can evolve toward the final setup. This process takes 1-3 months in most accounts, but experienced people can shortcut this delay. If the expert has done the industry before or has studied the competitors, they can often shorten the data gathering timeframe by 50% or more.

Account Restructure

The idea behind this strategy is to have budget controls on the three types of keywords. The only way to control the budget is by running your three types of keywords in separate campaigns. This structure results in a more complex account, but it is necessary.

Campaign 1: "A" Category Words

The first step is to create a new campaign. Separating each class of words gives finer control over the budget settings. Put the exact match "A" words into the new campaign and then the same keywords into a negative keyword list. Attach the negative word list to the B&C campaigns to prevent competing with yourself. You have enough enemies in the market, and you do not need to compete with yourself. The initial setup will give you an estimate for the budget on these keywords. Add an extra 20% on top of that. Create a new ad group for each word. The word separation by campaign allows you to fit the ad copy tightly. We typically see an improvement in the quality score, which is a bonus as these are typically expensive keywords.

Risk: During this time, you might have a budget overrun.

Depending on the traffic volume, the review period should be every other day up to a week. It is important to have enough data to make smart decisions.

Campaign 2: "B" Words

Create a "B" word campaign for the next layer of quality. For most businesses, this is where the budget runs out. If the budget runs out early in the day, then you must have the hard budget conversation. The business is losing profitable opportunities. Create a negative list of the "B" words and attach it to the A&C.

Campaign 3: "C" Words

Lastly, create a campaign for the remaining words. The purpose of this is to fish for more A & B words. This campaign is your experimentation budget. If this is not running out of money, then you did this wrong.

Managing Your New Configuration

Managing this set of campaigns is largely about budget control. You want to make sure that:

    A.   Never runs out of budget
    B.   Normally runs out of budget by mid-day
    C.   Normally runs out of budget

To simplify budget management, you can use a shared budget between A & B but **never attach C** to that shared budget. It is especially important to keep an eye on the impression shares at the campaign level and consider budget adjustments if A or B routinely runs out of money.

Impression Share

This tactic requires some thought and understanding of the impression share. Impression share is simply the percentage of times the ad appeared when it was eligible. When managing a constrained budget, this is a critical data point. You can lose impression share two ways, by Rank and by Budget. As the loss to budget shrinks, you need to watch the loss to rank.

A Few Rules

1. Never put broad or modified broad keywords in the A or B campaign.
2. Never let "A" run out of money.
3. Review once a month for words to promote or demote.
4. Treat C as a keyword testing ground.

By isolating the performance budgets, you gain greater control tightly connected to your ROI. There is an "Ignorance Tax" that gets charged to those that do not carefully track the details, and this strategy has lots of details. As is usually the case, you can cure the business problem with cash or creativity. Since **all** my clients have a limited amount of cash, we spend creativity first and recommend that you do the same.

Wow, Google Ads Do Work

People often tell us that Google Ads, aka Google AdWords, DO NOT WORK. They say that nobody clicks on the ads, it costs too much, and it just does not work. They have tried them, and that's that.

As a PPC Agency, we take a deep breath and reply, *"I accept the fact they did not work for you, but they do work."* That statement is followed by a colorful conversation that is always entertaining and enlightening. Let us explore the common issues people raise against Google Ads and see if they resonate with you.

Nobody clicks on the Ads

Nobody clicking is a self-healing issue because if nobody clicks on the ads, it costs you nothing; nothing ventured, nothing gained. Google earns $100 billion a year, and 96% of that comes from Google Ads, so if nobody clicks on the ads, why is Google so rich? Like most statements, there is a degree of truth in this one. Most experts agree that organic listings get 80% of the overall traffic. That makes sense since a significant percentage of searches are for research, and almost all those clicks go to organic listings. People click on the ads when they want to find a vendor to buy a product or engage a service, typically the traffic you want. To compete effectively, you must play both sides of Google. An organic content strategy for the research traffic and a PPC strategy for buyers.

It Costs Too Much

Sometimes true, but it usually means something is wrong. Some people look at business as a cost control mission, and the problem with this is that they can save their business out of existence. These people prioritize spending less when the real mission is to spend as much as you can while maintaining your CPA target. Google Ads should be managed from a maximum value, not a minimum cost perspective.

Google Ads can indeed cost too much, and when true, the fix is to use the tool properly. It is not fair to blame a screwdriver for not doing an excellent job driving nails; the same is true of Google Ads. Step one in this process is to calculate and understand your target CPA. If you achieve your target CPA, then you should **spend more**, not less.

We have seen situations where Google Ads have been the wrong tool for the job. When this happens, you must look at how the tool is being used. If Google Ads is not creating more value than it costs, you need to reengineer the campaign. Roll up your sleeves and get back to work.

Why pay for what we can get free?

When people say free traffic, what they mean is the organic listings created by your investment in SEO (Search Engine Optimization).

➢ SEO is only free if you have a source of free professional labor.
➢ You pay for SEO differently, but you still pay for it.

To earn organic traffic, you must have a steady supply of fresh content that applies to your business. There is a labor cost for planning, writing, optimizing, and publishing the content to be considered. Many owners have told me that this is their time, so it is zero cost. The reality is that they do not value their time, typically the most valuable time in the business. We have run hundreds of cost evaluations of free traffic, and most of the time, it turns out to be the most expensive of all the traffic.

Organic listings are the PR of the internet and its great traffic, but it is also highly competitive and ages poorly. Organic needs to be part of your Digital Marketing strategy, but it is exceedingly rare that a strategy can stand on a single tactic.

Organic provides extraordinarily little usable data. Analytics will give you the visitors for this traffic type but not the search query. This data leaves analysts guessing about what search query created the traffic. Google Ads and Search Console data can greatly increase your knowledge of organic performance.

We will just Do It Ourselves

The question is not can you, but should you? At first blush, Google Ads seems simple: throw in a few keywords, write a few ads, give it some money, and you're up and running. Google certainly pitches Ads that way, but it is just not true. Do the minimum setup, and you will surely get the minimum in return, if not less.

There is a performance component to consider. In the Military, soldiers are ranked on weapons as Expert, Sharpshooter, Marksman, and Unqualified. Ranking exists in all professions, and certainly, this exists in Google Ads. In most cases, what separates Experts from Unqualified is practice, and the same is true in Google Ads. Our staff works full time every day on their skills and most clients it is an additional duty given to the person with the most time on their hands. You can run Google ads with the unqualified, but do you want to do that?

There is an economic component to consider. If your time is valuable, then you should weigh the value you create in your business compared to the cost of outsourcing. In most cases, that value greatly exceeds what you would pay an agency. So yes, you can run it yourself. The question is, should you?

Keywords did not give me what we wanted

But you did get what you asked for. Keywords are not simple, and they certainly do not work the way you want them to. Poor Keyword Models are the number one cause of poor account performance. Simply throwing keywords at the system is an **accident waiting to happen**.

You will burn through your budget by buying non-responsive traffic that creates nothing for your business. 90% of the accounts we take over suffer from this problem. We develop the keyword model from the most specific highest intent value words and then loosen the keyword model until we reach the balance point between budget and market share. Simply stated, we want to buy the most targeted traffic right up to the budget. In a recent case, applying this strategy was a **performance increase of over 300%**.

A common criminal in this issue is broad match keywords, which truly are permission for Google to steal your money. These will surely give you traffic that is NOT what you were expecting.

The World Does Not Revolve Around You

Let us face it, most of us think the world revolves around us. In keywords, you must think about how your customer and the rest of the universe use your keywords. When taking over internally managed accounts, it is common to find features not benefits being highlighted. **Customers seek the benefits, not the features.**

For example, we had clients in the music licensing business, meaning they consulted on getting music rights for films. The challenge was that songwriters and musicians also **use these same words** for information about how they can license their music to projects. The client's business did not serve songwriters or musicians, but that traffic is most of the keyword 'music licensing.' In our terms, these are gray words, and to manage an account, you must realize that people can burn your budget out of your target. When managing these words, you must adjust bidding to bring the value in line with the cost.

It is PPC, so we don't care about Impressions

You could not be more WRONG! You should care about impressions, and here is why. In Google Ads, money comes in two forms; money, as we all understand it, and quality score. When your bid is processed, the ad rank is the money multiplied by quality. Stop and think for a second here – increasing your quality score reduces the amount of money you must spend for the same ad position.

Consider the following example:

Bid	Quality Score	Ad Rank	Win/Lose
$1.00	3	3	Loser
$0.43	7	3.01	Winner

Spend Quality or Money; it is your choice

What does this have to do with impressions? The largest part of the quality score is the click-through rate, which is clicks divided by impressions. Improve the CTR, and you drive quality score up. Higher quality score means that you spend less and get more. Quality score is Google's way of rewarding advertisers that pay attention to improving the SERP.

Quality Score

The Google agenda and yours are not always in sync, and CTR is an example of this. If you create a keyword list or ad copy that creates a lower CTR, then you are in conflict with Google's beliefs. Google believes, at its core, that CTR is an indicator of quality. In many presentations, we have seen them represent CTR as the **biggest** quality indicator. In most cases, a higher CTR is a good thing, but there are exceptions.

A client in the Web Design business came to us with a quality problem. Their Google quality score was above average, but as the discussion evolved, it became clear that the quality problem was the quality of the calls, not Google's. This business did high-end work with minimum projects starting well over $10,000. The agency on the account before us judged their work solely on Quality Score and based on that, they did a good job. The challenge is that 99% of the traffic volume for web designers is for projects well below this level.

To resolve this, we changed the headlines and included the $10k minimum project size. Both Click-through rate and quality scores dropped like rocks. By all Google measures and best practices, a horrible result, but the client loved it! They spent less and got more and better leads. Before clicking on the ad, prospects knew this was a high-end design solution.

Quality Score Math

Why do we care so much about Quality Score? Because of the impact on the cost of our clicks. The higher the quality score, the lower the bid can be for the same position. Let us look at a simple example:

Bid * Quality Score = Bid Score

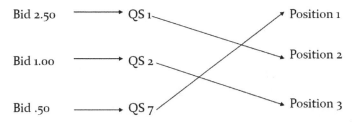

Quality score is closely related to the SEO score. It is an indicator that the site has an SEO problem in search as well. Some clients struggle to understand this, but Google values the "Quality of the SERP" more than MONEY! The reality of the search above is that the QS 1 would not even make it to the page.

PPC Bid Math

Bidding is a mystery to some, but it is one of the simpler parts of PPC. Let us do a quick walkthrough of this example:

Data	Description
$450	Sales Amount
50%	Gross Profit Percent
$225	Gross Profit Dollars
33%	Marketing Cost Allocation Percent
$74.25	CPA
14.0%	Conversion Rate
$10.40	Target CPC
20%	Auction Discount
$12.47	Average Bid
$24.95	Maximum Bid

Sales Amount is your average ticket, and $450 for an emergency plumbing call is a reasonable estimate. If you do not know this number, take last year's sales and divide it by the number of work tickets, and you have this number.

Gross Profit Percent is what's left after subtracting the ticket's direct costs of parts and labor. Sales times gross profit percent gives you the gross profit dollars.

Marketing Cost Allocation Percent is the answer to this question. How much of the gross profit would you spend to create the ticket?

CPA is the Gross Profit Dollars times the Marketing Cost Allocation.

Conversion Rate is the percentage of paid visitors that become a customer. For Plumbers in emergency keywords searches, this is normally in the 20% range. Then we must discount this by the losses in Customer Service and the Technician. Most Plumbers with well-run phone answering and good technician sales training will get .85 (15% loss) from the phone and .75 (25% loss) from the technician. The 14% shown here is the net of these. Every industry has different percents, but the process is the same.

Target CPC (Cost Per Click) is the CPA times Conversion Rate.

Auction Discount is the discount that happens in Google Ads. What you pay in Google Ads is one penny more than the competitor behind you. The math is fancier than that, but our goal here is to have you understand bidding. You do not need to replicate the math. 20% discount is a common number, but you can calculate this. Take the clicks you bought times your bid and actual CPC, and the difference is your auction discount.

The **Average Bid** is Target CPC plus the Auction Discount, what your Average CPC needs to hit to make the plan.

The **Maximum Bid** is twice the Average. That is the point at which you must back off the competition for the keyword. You can never get back to your average bid target if you exceed this number.

With Great Power Comes Great Responsibility

Google Ads are a powerful tool, and powerful tools deserve respect and care. At its core, Google Ads is just a tool, and when used properly, it can work wonders. Disrespect Google Ads, and it will set your credit card on fire with complete disregard for your financial safety.

Google is hiding information

Did you catch the notice in your account like the one below? Well, our friends over at Search Engine Land sure did and posted an article about vanishing Search Query Data. This article created a lively discussion in our Virtual Office. There is little doubt that Google will make this worse than it already is, but it isn't good. Google has been hiding search query data for years, and it looks like they are going to make this even worse. Not every advertiser realizes this, but there is an enormous difference between the clicks you paid for, and the search query total reported. With Google's first year-over-year decline in revenue, maybe it is time for advertisers to come together and bring balance back to this relationship.

> ℹ Changes to the search terms report
>
> We are updating the search terms report to only include terms that were searched by a significant number of users. As a result you may see fewer terms in your report going forward.
>
> LEARN MORE   DISMISS

Search Query Data has not accounted for the full advertising spend for as long as we can remember, and we started in Google Ads in 2003. In this latest release, the claim is that this is to protect privacy. To this claim, we must call BS and would like to know exactly how that works. Google has steadily stripped away control and detail in matching and reporting, and advertisers need to push back to get a fair balance in the relationship.

In the olden days, Google Analytics gave us organic search data, but that was stopped in the name of privacy, then moved over to Webmaster Tools, now called Search Console, where the data is, let's say, a little suspect. After 17 years of looking at Search Queries, we can tell you that whatever is in the Google Search Console is not search queries. We imagine this is rolled up data based on what Google thinks should be reported. Recently we have noticed that the percent of reported traffic in search queries within Google Ads has fallen from 20% to 10%, which makes the data worthless. Google is executing a strategy of taking data and control away from people and moving it to Machine Learning, where they can make sure that they always win.

Google is a Monopoly - So what?

A monopoly is not, by itself, a terrible thing. However, a monopoly is power, and absolute power corrupts absolutely. Google earned their monopoly, and they are entitled to reap the rewards of that. However, they need to act fairly and not use that power for evil. Google policy is largely fair, but not always. When it is not fair, there is no trusted, neutral third party to arbitrate the situation. We have watched Google go from private to public, and the policy slant more toward what is good for Google rather than what is fair. Do you remember the **"Do no evil"** days?

Advertisers Unite

Relationships only work if they are balanced, and that requires equalized power. The only way that is going to happen is if advertisers band together. No single advertiser can stand up to Google, and Google knows and acts like this. In 2012, Google made a change to the ad rotation, causing an uprising, proving they will back down when the pressure is high enough. The challenge is that they do this in little pieces to avoid poking the bear. If advertisers were united, they could be part of the decision process, making the process closer to fair. Advertisers have the right to demand that the money they invest can be tracked and understood.

Advertisers Should Have Options

In an honest relationship, when Google wants to implement a change, that should be fine, but advertisers should have the right to opt out. Google decides that measuring position is better with percent of placement rather than average position. This data is needed for the Machine Learning algorithm, but our account is managed better with the average position. We want the option. Here is a partial list of items that Google has taken away that should be advertisers' options:

1. Accelerated Delivery
2. Keyword Matching Control (Close variants versus strict matching rules)
3. Average Position
4. Placement Controls
5. Search Query Data
6. Ad Rotation (gave this back, but it is not a strict rule application)
7. Daily Budget Control (get rid of catch-up rules)

I am sure that with a little research, this list could be much longer. Just because Google likes it does not mean it's good for you, and I think advertisers should have an active role in policy.

Have You Lowered Your Bids?

Did you notice the drop in Google's stock recently? If you are like me and hold Google shares, it was painful to watch. The pandemic has changed advertiser's strategies, with some dropping out and changing

the CPC calculation. Year over year, revenue dropped for the first time in Google's history, supporting this theory of a shift to defensive mode. Since Google Ads are market-priced, a drop in competition creates a drop in cost per click. If you stand still at a time like this, you are spending too much.

Race to the Bottom

When you have the right conditions, the Race to the Bottom strategy is an excellent idea. But you must have the right conditions:

Lost Impression Share to Budget

Constrained Budget

Accurate measurement in place

There are tools you can use to reduce the bids, but my favorite is time slicing. It would be best if you did this in cycles of about three days between adjustments. This time is enough for the change to be in the data before the next round of changes. Impression share posting is commonly delayed, often by days. Google's documentation says it is posted several times a day, but our experience is one to three days. Watch the raw impressions for each cycle and then confirm this over a longer period with the lost to budget percent.

Reduce the bid, wait three days, then check conditions again. The increment of the bid reduction is a judgment call, but we normally start at 10% per cycle. When you hit bottom, then move the modifier up by 1% to do the fine-tuning. Race to the bottom requires continued adjustments, but it can produce impressive results. Remember that impressions are not consistent from day to day or week to week. When you hit bottom on Monday, likely, Tuesday may not. Daily you do want some budget left over because there is no such thing as a perfect setting.

Other Considerations

Once you have the bidding adjustments completed, you need to apply this to the keywords to pick up other opportunities. Suppose a 20% time adjustment gives you the number you want. Then look at the keyword performance and give the best performers more of the budget. This adjustment might be a 10% cut for one class of words and 30% for others, but this is the finer tuning.

Your Ego or Your Wallet?

Google Ads Bidding Strategy

In Pay Per Click advertising (PPC) systems like Google Ads, there is a trade-off between your ego and wallet, and it amazes me how often the ego wins. Most people are naturally competitive, and Google Ads plays this like a fine instrument. In the early days, Google provided the average position. Most people locked into "**Being in Position 1.**" Google changed that and now provides impression share, how often your ads show up, rather than position, but the game is still the same. We are having this conversation today because being number 1 is best for your ego, but a lower position can be better for your wallet. We strongly believe that the priority is to keep money in your wallet, not putting it in Google's.

Lower Position Strategy - Data Example

Let's look at what a lower than position one strategy looks like to your wallet.

Description	Ego	Wallet
Position	Top	Other
Budget	$1,000	$1,000
CPC	$5	$1
Net Visitors	200	1,000
Conversion Rate	2%	2%
Average Order	$200	$200
Total Revenue	$800	$4,000
Gross Profit	$400	$2,000
Marketing Expense	$1,000	$1,000
Profit/Loss	<$600>	$1,000

After a less competitive position, results went from a $600 loss to a $1,000 profit. Some conditions must be in place before you can use this tactic. There must be enough traffic that you can still spend the entire budget. You must believe that after they click on the ad, the ad position is no longer important. The targeting strategy must be changed from "Highly Aggressive" to "Conservative." although every market will be different. See the table under the Rule of Thumb section for these terms.

Does NOT apply to Everyone

There are situations where this is a bad strategy, and most have to do with emergency traffic. For example, a Plumber or Lawyer might have

Abs. Top strategy because once the person believes that the business will serve their emergency, they stop shopping. This strategy is for businesses with more available traffic than budget and who sell by making it to the shortlist. A shortlist is where the consumer sets the number of vendors they will consider. If a consumer has a shortlist of 5. Then first or fifth is the same because the choice is after the ad has served its purpose.

Go Ahead, Google – Talk Data to me!

There is no doubt that ad position matters. Many think the higher on the page, the better, but that is only one part of a multi-variant issue. Simply throwing money at your bid will improve the position but at what cost? If you have a budget constraint, then a higher bid reduces total traffic. To understand this, we need to examine the variables that follow:

1. Impressions
2. Clicks
3. CTR
4. Avg CPC
5. Search Impression Share
6. Search Abs. Top IS
7. Search Top IS
8. Search Lost IS (Rank)
9. Search Lost IS (Budget)
10. Search Lost Top IS

Let's get the acronyms out of the way.

CTR = Click Through Rate, Clicks divided by Impressions

CPC = Cost Per Click

Abs. = Absolute Top is the first Google Ads position

Top = Top positions above the organic and map positions

IS = Impression Share, total impressions divided by the estimated number of impressions you were eligible to receive

Rank = Loss from a low ad rank, bid times quality score

Budget = Loss from low budget

Reducing Bids = Reducing Shares & CTR

As you reduce your bids, your ads will slide down the page. Remember that ads only show on the top and bottom of the page. Important because the CTR drops steeply as your ads drop from top to bottom. It is not unusual to see the CTR drop by orders of magnitude from Top to Other. You can see this data using the "Top vs. Other" segment tool.

Google search: Top ⑦	3,389	269	7.94%
Google search: Other ⑦	2,449	17	0.69%

Contrary to what Google would have, you believe a lower CTR is not always a dreadful thing. If you can spend your entire budget at a lower CPC and CTR, that is good for your wallet. Assuming that your CPA (Cost Per Action) and VOA (Volume of Actions) stay in line with your goals, something you have to watch carefully.

It's the CPC Stupid!

We have borrowed from the famous phrase coined by James Carville in 1992. The purpose of the quip was to keep people focused on the critical issues, and it made Bill Clinton President. Here, we want you to focus on what drives your Google Ads. The Search Impression shares are all driven by the CPC, and it is your Bid, Quality Score, and Competition that sets your CPC. Your Bid is the one variable that you have immediate and direct control over. As you push the bid up or down, you will move the Impression Shares. The point is that the Bid is what you can change immediately.

Rule of Thumb

Different businesses operate at various levels in the market, and it relates to their competitive nature and target Impression Shares. The best one to work with is the Search Top IS. Here are the common targets for this number:

Strategy	Range
Highly Aggressive	75%+
Aggressive	50%-74%
Average	25% - 50%
Conservative	10% - 24%
Highly Conservative	Under 10%

As you move the bid higher, you shift more toward your ego and away from your wallet. Remember that you must watch the Search Impression share because that tells you if you are walking away from a market opportunity. You must watch the CTR because it is a big part of the Quality Score, another form of money. If your traffic is emergency traffic, then pay attention to the Search Abs. Top IS. If you decide on a strategy and use the table above in most cases, you will be close to your optimized balance point. From there, it will be fine adjustments with consideration of your CPA and VOA.

Budget or Rank

- Search Lost IS (Rank)
- Search Lost IS (Budget)

If you are losing Search Share to Budget, the fix is simple. Increase the budget and give Google more money. The loss to rank is a different thought process. When the Bid is too low to qualify for the impression, your ad does not show up. The fix is to increase the Bid, but that drives up the CPC and consumes the budget faster. At this point, you probably see just how complex this can get to be but wait. It gets worse. As you increase your bid to bring the rank in line, you will find that the last 10% is not worth going after because the increase in bid goes back to the first click.

If you are looking for one right answer - this is not the question.

Most sales and marketing challenges balance different variables, and variables are connected. Businesses always want to change one variable without having the others move, but that is just not reality. This specific challenge has an abundance of balance and string relationships. The key is understanding the relationships and finding the right balance for your business.

Bid Less Get More

Bidding more is good for Google but may not be the best for your business. We do not always agree with all the "Google Best Practices," this is one of those. Google makes impressions and sells clicks, so the CTR is their holy grail because it is how Google makes money. The reality is that you run Google Ads to help your business, not theirs. If you can reach your volume goals, then the goal is a lower cost. It is when you cannot reach the volume goals that you should consider increasing the bid.

Click Fraud

When Competitors click on your ads

Nobody likes to talk about this, but **Click Fraud does exist**. Google Ads is not perfect, but this is not as big a problem as some think. Google does a respectable job of detecting click fraud with automation and has spent billions protecting Google Ads from it. The days of accounts being impacted by a competitor burning through your budget by constantly clicking on your ad are long gone. Google technology cleans up and removes it from your account every day. This time delay can create problems when a competitor plays this game, and your account goes offline until the next check by automated monitoring.

The difference between good and bad traffic is the intent of the person on the keyboard. If they **have no intent** to use your services or buy your products but rather **intend to drain your budget,** that is click fraud. The challenge is that there is no "intent sensor." Much of the traffic is early-stage purchase research, and while it has no intent on this visit to buy your product or service, it can very easily be moving the prospect through your sales funnel.

Some propose that bounce rate is an indicator of click fraud, and this is sometimes true. The definition of a bounce is landing on one page and then leaving your website, which seems like bad traffic. However, we might change your opinion if we told you that a bounce could be good traffic. For example, a customer clicks on your ad, sees your phone number, calls, and buys your product or service. Homerun! **This example is also a bounce**, so be careful how you interpret your data.

Click Fraud is part of the cost of doing business. You should still monitor and work to reduce this but realize that the world is not perfect. There are many tools in Google Ads to protect your account, like IP address blocking or negative geographic targets, but most of this is a risk you just must accept.

Things You Can Do

Watch for extremely high click-through rates on individual keywords within the ad groups. Often the first clue that you might have someone playing games with your account. If your normal click-through rate (CTR) is in the 2-5% range, but one keyword is 30%, that word could be the one the vandal is using. If they do this too much, the Google tools will detect this and correct your account, but that might not be as fast as you are.

Watch your server log for IP addresses with unusual visits to your site and then exclude them from campaigns. IP blocking can quickly become a large-scale whack-a-mole game, but it is one of the tools you have. The challenge is that smart vandals will know how to spoof the IP address.

If you believe that it is a competitor, and you know where they are located. You can exclude that location, so they never see your ads. This tactic can costs you good traffic, so it is not the most elegant solution.

Watch for periods of abuse and exclude those. In many cases, we see that the vandals do the same activity at the same time, and you have tools to use for that.

Talk to Google

Talking to Google does not always help because most of the time, the Googler will tell you that everything is okay. We have noticed that the abuse seems to go away for a while when we have done this. We suspect but do not know that Google investigates and that causes people to behave better.

Advantages of Google Ads

Most will tell you that Google Ads cost too much, and we have never found an example where this was not true. However, compared to the alternatives, Google Ads is still a great bargain. In a world where a one-time ad in a local paper can cost $1000, and a full-page Yellow Pages ad can cost much more, Google Ads is less expensive AND performance-based. You only pay for the person that clicks on your ad and visits your website to learn about your business. In print media, you pay for space regardless of how much value it creates. If you get the right click from Google Ads, it creates value exceeding its cost.

➢ If the click creates a job – it is an investment
➢ If it fails to create a job – it is an expense and should be eliminated

It is not reasonable to expect that every click will create a reaction in your marketing, but the conversion rate must be acceptable to the business. You must be exceptionally careful with **what you think you know** when it comes to marketing data. In all cases, marketing measurement is inaccurate to some degree. The classic example of this is a person that does the following:

➢ Searches for a "Plumber Near Me" and clicks on your ad and scans your website.
➢ They remember your name.

- Later in the day, they search your business name and call your business.
- Do you credit this to organic or paid?

Not an easy question, but it happens every day. A customer will rarely respond to a single touch from your marketing. In many studies, consumer contacts to a business often had 20 or more marketing touches before contacting the business. This challenge becomes even more interesting when they shift between devices from their phone to their desktop.

Highly Targeted & Responsive

Google Ads can serve your ad to people searching for your services in your market. You can target individual cities, metros, or a radius around a location. You can even choose when to serve your ads, ensuring that they're only on when someone in the office can answer customer calls. The targeting and measuring the ability of Google Ads is well beyond any other form of advertising.

Plumbers get 25%-30% response rate

Pay as You Go

Let us face it you need the ability to increase or decrease your budget as your business changes. When you used the Yellow Pages, you were stuck with a fixed expense for a full year. In Google Ads, you can slow down or stop your advertising at any time. If you get booked for a large project, and you know you cannot accept new work, turn your ads off until you're ready to go again. We believe that keeping that money in your checking account is the best idea.

Google Ads - The Internet Yellow Pages

In the last generation, the Yellow Pages was a must-have advertising placement for home and professional services, but today it is Google. Google has replaced the Yellow Pages for most people, and this is becoming truer every day. The Yellow Pages aren't dead yet, but it is certainly not the "must-have" placement that it was several years ago. Dollar for dollar Google costs less and creates more value for your business. As the internet transitions to smartphones, Google Ads will be at the forefront of that revolution, putting one more nail in the coffin of the Yellow Pages.

Disadvantages of Google Ads

Every type of advertising has its negatives, and this is true for Google Ads as well. Here are the big ones:

Google Ads is Self-Service Advertising

Google Ads is a self-service advertising system that is easy to understand. Suppose you happen to be a programmer with advanced skills in keyword modeling and data analytics. At first glance, Google Ads looks simple, pick keywords, write some ads, and you are off and running. Simple is what Google wants you to think. The reality is that Google Ads is a competitive marketplace with hundreds of variables that you can configure to your advantage or disadvantage. There is a myth that you can set Google Ads up once and just let it run, and that is an accident waiting to happen. In the fiercely competitive area of Google Ads, we can assure you that a **"set it and forget it"** strategy will not work. Your competitors are not managing dormant accounts, so you would be standing in place while they beat you senseless, which is not in your best interest.

It is Expensive

When professionally managed, Google Ads can be cost-effective, but they are never cheap. This is especially true with the major keywords of a service name and the city. These are the money words for local services, and everyone knows it, so the competition is tough. Google Ads is an auction, and so it is the competitors that set the price per click, not Google. Competition can get out of control when you have competitors that want to be top of the page regardless of cost. The best way to respond is to hold steady and wait for the economics of their business to adjust their outlook.

It is complicated

Google Ads only looks simple. As you dig into the details, you quickly realize that it is a complicated text modeling and competitive bidding system that can become incredibly complex very quickly. The math and language issues involved in Search engines are far more complex than most modern programming languages, and it takes a unique set of skills to master this. Google Ads requires systems expertise to get maximum value from the investments being made. Google Ads can mislead you because it is amazingly easy to run Google Ads, and with a single account, it isn't easy to know if you are doing okay.

It is rapidly changing

In the World of Google, change is the one constant. Google Ads has a steady stream of changes, and the pace only seems to get faster. Google averages more than one change per day. Some changes are big, and some small but daily change is normal. We have worked in Google Ads since 2003, and we have seen three complete system replacements and untold smaller changes. A Google Ads Expert 3 years ago would be completely out of date today. It is common to see hundreds of systems changes each year.

Branding

Let us start by explaining why branding your business is important to controlling your costs in Google Ads. When we build a brand in a market, traffic starts to build not on the generic term of "Service Name" + City Name" but rather on your unique name. In the age of the Yellow Pages, the goal of branding was to get the consumer to look your business up in the White Pages (by Name) rather than the Yellow Pages (By Business Type). This tactic is powerful because they do not see all your competitors at the point of purchase. When they search on your brand, the cost of the click drops, and the likelihood of getting the traffic on an organic listing increases. Branding is done the same way on the internet that it is done in the physical world. You place your brand and message in places that your prospects are likely to see. These are placements like radio and TV stations, local papers, and other common community placements. These impressions are no different than branding your trucks or banners at the local baseball stadium. Keep repeating your brand until it becomes what people think of when they need your services.

Branding is critical to building a successful business; it is also one of the most misunderstood parts of a business. Many people think that branding is just a logo or slogan, but it is much more than that. Branding is built by all the communications between the business and the market. The brand is literally what your business comes to mean in the market.

Branding is just as important for a small business as it is for a Fortune 500 company. Brands can be tag lines, logos, or other marks representing your business and are uniquely yours.

Branding is well beyond the scope of this book, but Google Ads should be part of the overall brand. It extends the reach and frequency of the brand. In branding, coordinating and supporting the overall message is key to impression strength, and text-based ads are not strong brand impressions. With its image ads, the Google Display Network is a great tool for creating quality branding impressions. Display ads are especially powerful when used with remarketing.

Since Google Ads can support direct response and branding, one inescapable issue is finding the right balance. In most SMB's this means balancing the long term (branding) with the short term (direct response). We commonly recommend to clients a simple percentage distribution of 10-20% for branding, with the bulk of the budget going directly to response ads. Branding is how you win the long game, but you need most of the money flowing to the short term so you can afford to invest in the long term.

The designer should know the goal is not to get a click - the goal is to make a brand impression. We want them to remember the brand and search for it later. The performance measure for these ads is an increase in organic searches for the brand. We commonly will set up a search campaign to get brand volume data to measure these efforts. Measurement is not perfect, but it is better than the complete darkness we had before digital.

Generally, this means an ad design with the brand name and a clear statement of what the brand is about. You only get one tagline for a brand, so pick it carefully. Never confuse the function of a response ad with a branding advertisement. A branding advertisement does not have a call to action, although contact information is fine.

Display PPC Networks

Display Networks are advertising placements based on the person visiting a site rather than the search they conducted. Google's Display Network (GDN) is estimated to be over two million sites. Display Ads are less effective than a search because of the lower intent of the audience. In search, the person asked for information; in a display ad, they ran into it. The degree of this depends on why the person is looking at the content you targeted.

Why Run Display Campaigns

Top 10 Google Display Network Campaign Ideas

The Google Display Network is arguably the most powerful tool on your marketing toolbelt. These are ads that your prospective customers see as they browse the internet instead of the ones they ask for in search. This ad delivery type changes the user intent factor. With careful planning, display ads can craft a branding image in the minds of your prospective customers. Seeing your ad on a trusted site can also cause goodwill to transfer to your brand.

Are you looking for a breakaway sales strategy that unleashes your business potential? Then get yourself a cup of coffee and read this section carefully. Here are the top 10 reasons to use Google's Display Network:

1. Driving search volume
2. Branding
3. Innovation Introductions
4. Targeting Strategies
5. Remarketing
6. Content Testing
7. Reduced Costs
8. Early sales Cycle Contact
9. Advertising Copy Creativity
10. Thank Before you Bank

Driving Search Volume

If a customer is not actively searching for you or your services, how are they supposed to know about you? The answer is: utilize a Display campaign to reach this potentially interested audience. Getting your ads in front of a customer that has yet to discover you is the first step in generating that qualified search to your website. Simply put, display ads allow you to place your ad in front of a customer that has never heard of you or your product. In time this leads to future searches directly to your site. Display ads are a marathon, and search is a sprint. You need to have patience and a steady budget to reach the finish line in display and not burn out before your opponents.

Branding

To move searches to your business from a generic term such as "Plumber" versus a branded term like "Joe's Plumbing" is impossible if you are relying solely on a Search Campaign. Display ads offer the opportunity to bring your specific business goals and values to life. You can highlight the unique selling proposition of "Joe's Plumbing" versus the other plumbers that might pop up in a generic Google search. For instance, a brightly colored ad with a memorable slogan about Joe's Plumbing will linger in a customer's mind, so the next time they need service, they have a predisposition to call Joe whether they are aware of it or not.

Innovation Introductions

Consumers will not search for something they have no idea exists. Innovative ideas, products, and services can be easily highlighted in a display ad. An introduction to what you offer is a unique opportunity that display-ad campaigns present. Suppose you are introducing a new robot that walks your dog. It's unlikely that anyone will search directly for this because they have no clue that it exists. Display ads are perfect for this, especially on sites featuring the latest tech gadgets. Someone browsing the web may see your ad and realize that they could use some assistance with their rambunctious rescue dog. The ad presents a solution the consumer is unaware of, and it is unlikely they would have searched for this.

Targeting Strategies

It may seem those display ads are akin to taking a shotgun approach to the market. However, the wide range of targeting options that Google offers can narrow down the people that see your display ads. In addition to targeting users who have expressed intent (determined by Google based on user activity), you can choose to place your ads on reputable websites that you know your potential customers visit. This approach is valuable because it provides a soft push to people whom you know already have a qualified interest in what you offer. For example, you can choose to target your ad to people interested in sports, even when they are not browsing a sports-related website. These audiences are not perfect, but they provide a great starting point for generating that initial "buzz" about your business.

Remarketing

Have you ever looked at a camera on Amazon, and then every place you go, Camera ads seem to follow you? That's remarketing, and it works incredibly well. Remarketing results in a much higher likelihood of engagement on your site than other targeting options. A remarketing audience allows you to show a message to those who have already visited your site. Let's say someone got to your site and didn't set up an appointment on your "Contact Us" page. With Display, you can show a specific and compelling message in your ads to this group of users. The result is a higher probability of that person coming back and exploring your website. It is very unlikely for someone to do business with you on the first visit making remarketing a powerful tool. It gives you a second chance, which we know isn't common in advertising.

Content Testing

Due to the increased number of viewers of your ad, display ads offer the chance to **fail fast**. You can quickly evaluate whether a campaign produces desirable results and moves on, whereas, in a search campaign, this process would take much longer and cost more money. It is critical to know what content works and what doesn't. Many times, content decisions are based on the personal choice of the highest-ranking person in the room. It goes like this: three ideas are presented in a meeting, and the CEO likes option 2. The result is that option two is picked even though the CEO may not fall in the target audience. With display, you can put the three concepts into a test and let the market tell you which one works best for the real audience.

Reduced Costs

Typically, display campaigns produce a lower CPA with a longer sales cycle. The key to managing this is to keep your bidding finely tuned. The more you can fine-tune your targeting, the more cost-effective your display campaign will be. If you tighten targeting in your display ads, the cost per lead becomes much cheaper than search. Therefore, display ads have a lower cost, especially when considering the quality of the audience.

Early Contact in the Sales Cycle

Unlike search driven by the user's query, Display is driven by your detailed reading habits of your ideal customer. Display ads allow you to reach your ideal customers early in the sales cycle and bring them through a full experience with your business. If you can understand, empathize with, and communicate to your ideal customer, then display ads can be a breakaway sales strategy that unleashes your business potential. Search targets users late in the purchase cycle, while display ads target users earlier.

Advertising Copy Creativity

The chance to unleash your creativity in display ads is endless. In search, you are limited to text which does not elicit emotion as an image can. Display ads make it easy to use shocking statements or humor to connect with your audience on a higher emotional level. While this can be risky, it can also be extremely rewarding for connecting with a specific audience of the people you want to reach with your message. In addition, the opportunity to create engaging ads in a variety of formats lets you give people a feel for your business and even your personality.

Thank Before you Bank

Would you like your customers to remain delighted customers? Would you like them to refer your business to a friend? Saying "Thank You" in a display ad helps promote a valuable connection that keeps people coming back. This simple step is often overlooked as it is easy to focus on acquiring more leads. However, etiquette and customer service go a long way in fostering relationships with people. Specifically, these efforts are likely to result in a referral from your delighted customers because you will stand out from those who did not take the time to say "Thank you."

There Google's Display Network is all about **reach and frequency.** Although reach and frequency are difficult to control, display advertising has a massive reach which lends itself to many opportunities to adjust your frequency based on your budget.

Managing Google Display Network

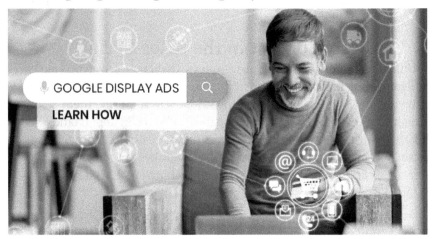

What if we told you that there is a strategy to scale your business beyond your most aggressive business sales plan? Would you want me to keep talking? If yes, then join us for a discovery of the Google Display Network. This chapter will explore **how to** manage the Display Network and explain **why to** use this tool. This chapter explores the basics of taking care of the display ad campaigns and how to create breakaway sales opportunities.

Display ad targeting is done with keywords, audiences, and direct placements. The Display Network is significantly less expensive than Search on a per-click basis and typically lower quality traffic. The response tends to be lower, so the resulting CPA is often remarkably close between the networks.

Here are the top things you need to tend to in your Display Campaign.

Minimum Monthly Maintenance Items

1. Placement Strategies & Reviews
2. Bid Adjustments
3. Advertising Strategies & Test Reviews
4. Budget and Impression Share
5. Google Analytics to Google Ads Verification
6. Competitor Strategies & Tactics
7. Waste Management

8. Analysis & Reporting
9. Outsourcing

Placement Strategies & Reviews

In Display, the three most important items are location, location, and location. Display Targeting tools include keywords, audiences, and placements. Display ads are especially important to innovations where search traffic does not exist because the market is unaware of the solution you are bringing to the market. When they do not know enough about you to find you, then Display is your tool of choice.

Audiences are lists of people from various sources with a wide range of quality. One of my favorite audiences is a remarketing audience made up of people that visited specific pages on your website. Audiences, demographics, and topics are largely the same thing sliced in different ways. They are groups of people gathered by different methods. There is little doubt that Google likes audiences. We just wish they were more accurate in audience selection and more transparent with the sources and rules.

Keywords are used to match the content of the page. When you dig into the details, keywords can be sloppy, with the match being downright creative. The upside of keyword matching is that it often gets lots of traffic, but you often pay for that with a poor match.

Managed Placements are the most accurate and the most work. Google has, over the years, made the use of managed placements more and more difficult with strong recommendations to audiences. We must be careful with the wording here because placement is where your ad showed up, while managed placement is a targeting method. Technically the terms are Managed Placements for your targets, but the managed word is often dropped, leading to language confusion.

One of my favorite strategies is the use of audiences and keywords to find sites that perform well. Then push those discovered sites into Managed Placements. This tactic gives us the wide reach of the audiences and keywords and the precise control of managed placements. As this strategy matures, the targeting becomes more precise as we move the budget from the fishing (keywords and audiences) to Managed Placement.

Bid Adjustments

There are many ways to create a bidding strategy for a Display Campaign, and the details could fill an entire book. We look at major placement factors:

- Quality of the Site
- Competition for the placement
- Content to influence factors
- Readers state of mind
- Past performance
- CPA Goal
- VOA Goal

Sites come in a wide range of quality levels, from premier placements to complete junk, and this is relative to the brand strategy and values. Some of my favorite placements are local media sites like Newspapers and Magazines, but sometimes the goal is the raw volume where sites like weather shine. Like all the Google Ads systems, the cost is market-driven. It is your competitors, not Google, that is setting the value. The content they are reading drives the audience's mindset, so finding a relationship between your ad and the content is important.

Advertising Strategies & Test Reviews

Strategies in Display are as creative as you are. This task is simply going back and reviewing your assumptions and projections. Explore your targeting method, and goal then compare to what you got. It is common to aim for one group and end up with something entirely different. You target businesses and then get traffic from kid's game sites. It happens, and you must constantly adjust your targeting to get back on target. Display targeting is more like a shotgun than a rifle, and it requires constant attention.

Display REQUIRES testing, and on a regular schedule, you need to check in on those tests. Display Ads are a general population tool, and different parts of that audience will respond to different rules of influence. You need advertising-creative for several of the weapons of influence. Your test is not measuring just ad to ad but also influence to influence.

Budget, Bid, and Impression Share

There is always more advertising available to buy than there is budget to buy it with.

With this as a given, the question becomes, "How much of the share do you want to buy?" The answer depends on the CPA (Cost Per Action) and VOA (Volume of Actions) goals. If your VOA is short and you have room on the CPA, increasing the bid is a promising idea.

As you buy deeper into the impression share, be careful about your costs. In this example, suppose we push the bid up from $.50 to $1 to get the last 10% of the traffic. That bid adjustment applies back to the first

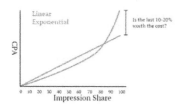

click, driving up the cost per click for ALL the traffic, not just the last 10%. This bidding changes the mix of placements you are purchasing since you now qualify for placements you lost with the lower bid.

Google Analytics to Google Ads Verification

Like all advertising, you are throwing money around, and auditing is important to maintaining control. Our recommendation is simple. Once a month, make sure that the traffic you are paying for is showing up. Look at the clicks from your ads and the arrivals on your page and make sure most of it makes it to the site. Web traffic has known tracking flaws, so the numbers will not match perfectly, but if the difference is more than 10%, then start digging because something is probably wrong.

Competitor Strategies & Tactics

Competitors are sneaky, evil, smart, dangerous, alpha predators in constant motion. They watch what you are doing and develop counteractions. They drive your advertising costs up, and they are not going to roll over and let you have the market. You are in Word War with your competitors, and it's serious. To play on the Display Network battlefield, you need to be the smartest of the alpha predators, requiring constant innovation. Stand in one place too long, and your competitors will figure out what you are doing and beat you senseless. Develop a constantly changing set of tactics, so you become unpredictable and unstoppable. Every month, you need new tactics and new targets. Find new groups that your product or service fits and find an emotional button you can push.

Waste Management

The Display Network is vast and messy. Add to that the fog of a Word War, and you need waste management. You will never make the system perfect, but you need to make it better than your competitors, or it will slowly bleed you out. Watch your placements and measure everything. If a site starts to display injurious behavior, you must act quickly. If you are not removing 10% of the site placements each month, you are probably not doing the waste management you need. Your strategies and tactics need to replace the volume you are removing, so this works hand in hand with the section above.

Analysis & Reporting

Management rule 1.01 states, **"Measure and it shall improve,"** and this is true. At the top end of analysis and reporting is the CPA and VOA, but the devil is in the details. Ensure that those who need analysis/reporting get enough detail to correct the tactic or strategy. Make sure that your analysis is looking to the future as well as the past. Reporting can be tricky to get right because you do not want to bury managers in data and starve them for information, and you would be amazed at how good digital marketing is at doing this. If the report is going to a manager who controls strategy, then stay out of the painful details they do not act on. The key takeaway here is to make sure the reporting matches the needs of the reader.

Outsourcing

To outsource or not to outsource, that is the question, and it's not an easy one. This chapter is written for two types of readers, what we call the DIY (Do It Yourself) and DIFM (Do It For Me) audience. For the DIY, this is a list of items they need to include in their work plan. For the DIFM, this helps them understand the work we perform for our clients in this area.

Many executives overlook the overhead of staying up to date in a rapidly changing specialty. Digital marketing changes at warp speed, and Google normally has at least one change every day. Google Experts typically spend 8-12 hours a week to keep up with changes. Internal staff typically manage one account, so 100% of this cost goes there. Agency staffs manage up to 30 accounts, greatly increasing the leverage of this fixed expense. There is great leverage in working on a series of accounts on the same task. Agency staff has peers they can learn from and collaborate with.

To be fair, the internal staff has its advantages. They understand the culture of the business and should feel a stronger relationship to the organizational goals. They have direct relationships with other departments and will know the current events of the business.

Staff turnover

When internal staff turns over, this often means losing 100% of the account management skills and tribal knowledge. The new person hired to replace the expert will often want to completely re-engineer the account because they do not understand the strategies and tactics. Agencies face turnover, but since they typically have multiple people with these skills, turnover results in overtime, not a complete system failure.

The Google Display Network is a tool that provides great power to the business to reach people not actively seeking their product or service. Display Ads are a great tool for new business and brand-building opportunities. It is not a magic bullet; it has its challenges, and it requires smart people that can use the tool to create an advantage for your business.

Display Network Direct Placement

Would you rather be a big fish in a small pond or a small fish in a big pond? GDN is estimated to be more than two million websites, and there are lots of places you want to be and lots you do NOT want. Display Ads give us several ways to target, with the big ones being keywords, audiences, and placements. Each placement type is useful with different degrees of control. We are control freaks, so our first choice in GDN is always Direct Placement. Keywords and Audiences are a great way to go fishing if you need raw volume.

Anyone who has worked with GDN will know that keywords are at best loosely controlled, and audiences are even worse. They are both extremely easy to implement and maintain. Placements, on the other hand, are tightly controlled with more work to implement and maintain.

Reach and Frequency

Advertising is driven by reach and frequency. If budget is a constraint, which it always is, these items must be balanced against each other. The bigger your reach, the lower your frequency, assuming the budget is a constant. So big fish in a small pond or small fish in a big pond?

Google Does Not Like Placement

Why Google dislikes placement is debatable but that they prefer keywords and audiences is not. Just go through the editor and examine how difficult they made researching the sites. They do not like giving the advertisers control, and **coincidentally that is the reason we love it.** Keywords and audiences are sloppy, and Google always wins when you give up control.

Research is Key

The key to Placement is to find sites where the readers will be in the right frame of mind to listen to your message. You want sites in the GDN that contain content related to your product messaging.

Stepping through the Bidding

Once you have identified the sites you want your advertising on, the next step is to figure out the placement cost and decide if that price works for you. Since the quality of the traffic is less than the search network, you must find the market cost. We do this by starting with a base bid then let the system run for a few days. If we get zero traffic, we increment the bid up by 25% and try again. Once the site starts to show traffic, then we stop incrementing that site.

Maintenance

Once you get the GDN campaign running, it must be maintained. Bidding will change, and new sites will join the GDN program. We recommend monthly bid reviews and a quarterly update on-site research. We have found over the years that this is a reasonable balance between cost and performance.

Automation

As you might have guessed, this process is detailed work, and we have invested in automation to help us. This software takes our target customer and models their website content to the placements we have already modeled.

Search Engine Optimization

SEO is an interesting field of study that has evolved since the early days of the internet. Search Engines seek to provide the searcher with the best results to satisfy their query. All businesses want to show up every time the search is in any way related to their business. It is the conflict between two goals that makes SEO fun.

SEO is a game of balance between the numerous factors that search engines consider. No search engine is going to give your business a win every time. You just are not that important. That does not mean that you cannot win your fair share.

> ## "SEO is done when Google quits changing things, and all your competitors die"

To SEO not to SEO?

hat is the question Whether 'tis nobler to suffer, the slings and arrows of SEO or be loyal to a rich reader experience is a fundamental strategic decision in any Word War. We are having some fun with the Shakespearean language, but this decision is foundational, complicated, and critical. All strategy decisions are a balance between conflicting goals, and this is no exception. On-page optimization is about keywords, densities, page rank, and the match to the user's search query. On the other end is the reader's experience, often compromised in the name of optimization.

Order of the SERP

Businesses must consider what they get for their optimization efforts and where that ROI comes from in this strategic decision. If the organic position in the SERP (Search Engine Results Page) is the upside, you must consider that much of this is below the fold (first visible page section). Being below the fold means the searcher must roll the page down to see it. While studies vary, a general rule is that 50% of the traffic is lost with every interaction the searcher has with the results. Consider that this is after the Guaranteed Service Ads, Google Ads, and Map Listings.

In most cases, if the only return were the organic listings, we would not even compete for it because our studies have consistently proven this to be the most expensive position on the page. The top positions of the SERP contain many impossible to beat sites, including Angie's List, Home Advisor, and Yelp. This reality puts the first truly competitive organic position on page two, the place where websites go to die.

Not the only factor

The organic listing is **not the only return** of your optimization investment. Consider that your search optimization score and Google Ads Quality Score are closely related. Improve your optimization, and improve your quality score. Over the years, we have seen SEO oversold, and it makes us crazy. It is pitched to businesses as the sole solution to their traffic needs, which is not true. It needs to be part of the overall strategy, but we have never, in 26 years, seen it succeed by itself in local service businesses. Optimization has a role in the map, although it is a secondary item. If all other factors are equal, then your organic score will affect the map results. SEO optimization is not the only consideration for maps, but it one of them.

SEO is an art form that balances the reader content experience with keyword optimization. We advocate for a strong **lean toward the reader experience,** with the last step being optimization. We reason that it makes no sense to get traffic to a page that does not create a **great reader experience**. What never works are attempts to optimize without a content strategy. If what you are trying to optimize is just your website content, save your money and take your significant other out for a delicious meal.

Over the years, we have seen some SEO success stories and far too many spectacular crash-and-burn stories. What separates winners from losers is the passion of the writer and the care taken in crafting the reader experience. The truth is that many SEO success stories are accidents created by passionate people that value what their business brings to the market. For SMBs to be one of the rare success stories, they must start with a content strategy that educates their market on the value they create. Listen carefully to the questions your customers ask and then tell the world. Your daily customer interaction is a huge advantage, and you should use that.

Order of Battle

SEO is an important part of your web strategy, but it comes after the web experience and initial PPC implementation. The web experience priority is self-evident. PPC goes first because it starts fast, and SEO often takes 6-18 months. PPC provides critical information needed to focus the SEO efforts. Without PPC's data, SEO can be like shooting with your eyes shut.

THE ART OF WORD WAR TACTIC

Learn
How to be
1st on Google

If you want to be first on Google organic results, you must EARN it. Some will tell you that Organic AKA Search Engine Optimization (SEO) is fast, free, and easy. All you must do is pay them to learn the secret. Unfortunately, Search Engine Optimization is none of those things, and the person selling the Search Engine Optimization secrets is a fraud. Search Engine Optimization is and has always been earned by creating high-quality, engaging content that communicates the value of your business. This chapter is an exploration of Search Engine Optimization in plain business English with no hidden secrets.

In Search Engine Optimization, a page gets points from the pages that point to that page. This is Page-Rank, which is the core of what makes Google, well, Google. Pages that point to your page transfer a portion of their Page Rank to your page. Then your page passes that on to the pages it points to. Not all inbound links are created equal, and here is a visual of how this works. Page Rank is simply the proportional sum of the page ranks of the pages pointing to your page. If your homepage has a rank of 25 and points out to 5 subpages, each subpage will get 5 points. The Page Rank is then adjusted by the

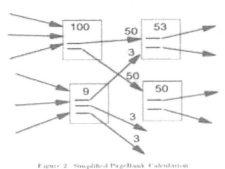

Figure 2. Simplified PageRank Calculation

image Credit: http://ilpubs.stanford.edu/808/422/1/1999-66.pdf This page includes a detailed discussion of page rank and is a great resource

density of the keyword on the page. Keyword density is the weighted occurrences divided by the total weighted words. Position on the page matters, and our best guess of the hierarchy goes like this:

Earn Points

Earn Rank
Earning Points

1. Title (20)
2. Meta Description (10)
3. H1 (9)
4. H2 (7)
5. H3-H6 (5)
6. Anchor text (4)
7. Lists (3)
8. Body Text (1)

The numbers on the list above are our best estimate of the weight of the position. Nobody outside of Google knows the real numbers, but this is remarkably close to the observed weight given in search results. What nobody disagrees with is the Page Rank and Keyword Density are foundational concepts of Search Engine Optimization.

Home Page Optimization

Optimization starts at the homepage. It is here that you first learn about Search Engine Optimization compromise and balance. The first rule of Search Engine Optimization is that **"A web page can only be optimized for one keyword."** Nobody is ever happy with that, but that is the rule. For all SMBs, the most you can compete for with your homepage is your business name and one keyword. This not magic; it is Simple Search Engine Optimization math. The keyword density distributes your Page Rank. Density is the keyword's weighted occurrences divided by total page weighted occurrences—the more words, the lower the density. In a simple example, if the page had 10 points and the keyword had a density of 10%, then the page would get 1 point. Nobody knows the exact math, but most experts will agree that optimization works like this. If you have those 10 points and compete for one keyword, you get 1 point. Compete for two, and you get 0.5. Every keyword you add to the targeting makes your keyword rank lower. Question one in optimization is, **"What is your homepage keyword?"**

All the site pages are optimized. Now what?

Once all your website pages are optimized, then ask yourself, **"Are you satisfied with the results?"** If the answer is yes, you are not cut out for marketing because you should never be satisfied. The reality is that optimizing site pages is just the first few steps of an exceptionally long journey. Optimization is a process, not a project, and the work is never done. If you want to win the optimization battle, you must take it to the next level.

What SEO is NOT

The above items are corrections to gross mistakes made on critical pages, not SEO. These mistakes cause all sorts of collateral damage but fixing them does not make your site competitive for organic traffic. When a business wants to optimize its site, the goal is always the same. It is to **gain a better position in the organic results.** Fixing the problems above might move you from not on the page to eligible, but it will do nothing for your rank. Competing for an organic position requires a steady stream of related, high-quality content placed on your site over a prolonged period. The level of this content must be greater than your best competitors.

Setting Expectations

If what you are looking for is a tactic that will move the needle on your sales in the next 60 days, SEO is not what you need. Winning a competitive SEO position in an existing market is a 12-24-month effort, and in many cases, it can be even longer. The rest of this discusses how to compete for those highly coveted organic positions. The competition for these is tough, and you must win and then defend your positions. True SEO is a process, not a project, so you must commit to an indefinite term.

The homepage is not the last time you will fight with this concept, but the reality is that **"The more things a page is about, the less it is about any of those things."** After the homepage optimization, the next priority is the directly connected pages. Each of these pages will get a proportion of the homepage points. The challenge is that the more pages you attach to the homepage, the fewer the points each one gets. You have to question the value some pages get, like the contact page. If you applied the Grunt Test to your site, then the contact information will be on the homepage or, better yet, in the page template. If that is the case, then **the contact page is duplicate content**. Keyword Density is a percentage, so it can be moved by changing either the numerator or denominator. Increase the weight of the keyword or reduce the total weight of all keywords. If you have lots of pages connected to the homepage, this reduces the weight that they pass to the next level of pages in the site hierarchy. Keyword density is a sweet spot attribute, and overdoing this can take you from higher density to a penalty for keyword stuffing. Once you hit the optimal point, higher density becomes negative.

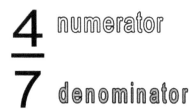

$$\frac{4}{7} \begin{array}{l} \text{numerator} \\ \\ \text{denominator} \end{array}$$

One Word Per Page Means One Word Per Page

Over the years, we have only seen one strategy that wins the long game of Search Engine Optimization. It is a steady flow of high-quality content targeted to common questions from customers written by passionate people. In your Google Ads account, you will find topic clues in the search terms. These are the searches that your visitors entered before they saw your ad. If you spent that money on purpose, then clearly that was a search you value. This search term is the core of the question they asked. Your content tactic should be to answer these questions and win the traffic away from your Google Ads. The longer your account has been running, the more valuable this data will be to your editorial plan.

SEO Editorial Policy

Once you have decided to compete for organic traffic, you will need a policy and calendar. An Editorial Policy is how much of your content will be editorial and how much advertising. Editorial content is **why people**  is come **to the site,** and advertising is **how content gets paid for**. These are symbiotic because without editorial, you have no traffic, and you have no business without traffic.

Set the Search Engine Optimization Pace

Search Engine Optimization is a long game, and you need to set a pace you can live with forever. In most SMBs, this is 1-2 articles per month. In addition, you need a modest ad budget to bring in new readers and an email list to remind current customers and readers that you still exist. The actual pace you need to set is faster than your competitors.

SEO Editorial Calendar

To create your Editorial Calendar, calculate your article slots. Then organize article ideas and put them in a priority sequence. If you are running the Search Network in Google Ads, turn to that data to help prioritize. Take the impressions and adjust them by the impression share and multiply that by your average CPC. This calculation will give you the max value for the period. You will never earn that much with your article, but it is important compared to the other articles. Create your calendar and review it each month and modify it based on market events.

Analysis & Reporting

Like all business processes, Search Engine Optimization should be on cycle analysis and reporting. For Search Engine Optimization, we use a monthly cycle and examine micro and macro data. The micro-data is the Search Engine Optimization channel in isolation, and macro-data is the company-level data. Search Engine Optimization traffic is often early-stage traffic in the sales pipeline. Connecting early-stage traffic to sales transactions is fraught with technical challenges. The Search Engine Optimization channel should be measured at the CPA (Cost Per Action) and VOA (Volume of Actions) for goals and actuals. Here is what that might look like:

Search Engine Optimization Channel

Description	Goal	Current Month	Prior 1	Prior 2
CPA	$25.00	$22.95	$26.95	$28.45
VOA	10	9	11	12

GDN for Search Engine Optimization

The Google Display Network (GDN) is a high-volume, low-cost traffic source, which is ideal for priming the newly optimized page. GDN starts with a placement strategy that can be site-specific, keyword, or audience targeted. We are fans of site targeting, called placements, in campaign controls. Where you specifically target websites by their name. This Display Ad targeting is the most exact control in the network. Keywords are the least controlled, but they are good for lots of traffic volume. Audiences vary in quality depending on how they were set up. The best is normally the remarketing audience built from prior visits to your website. Industry audiences built and maintained by Google vary widely in quality and need to be evaluated carefully. The simple idea here is to pick up early-stage traffic and get your editorial content in front of them.

Email for Search Engine Optimization

Building and maintaining an email list is essential for good Search Engine Optimization. A strong email list gives you a way to push your updated content out proactively and organically to your current customers and prospects. While it takes a long time to build a good opt-in email list, it can be your most powerful marketing tool. Our advice here is to get started on this today and never stop growing your list.

# Additional Reading

Search Engine Optimization is a complex topic that takes thousands of hours to reach a mastery level. Search Engine Optimization changes rapidly, so by the time you reach a master level, Google will change all the rules. To learn Search Engine Optimization, you need to get opinions from multiple sources and weigh the conflicts between these professionals. Over the years, we have come to respect the work of Search Engine Land and Bruce Clay. Search Engine Land publishes a great Essential Guide to Search Engine Optimization and a Periodic Table of Search Engine Optimization Factors, both of which are excellent resources. Bruce Clay is a competitor whose work we have grown to respect. We recommend his blog if you want to get into the details of Search Engine Optimization. Bruce Clay has been working in Search Engine Optimization since 1996, and we started in 1994, so we are both in the Internet Pioneer Generation. Bruce's strategy has been SEO first and PPC second, while ours has been the other way around.

Who You DON'T want to listen to?

The Search Engine Optimization industry has lots of criminals pitching the trick of the day or the secret to Search Engine Optimization. Google always figures out the trick of the day and comes down hard on those that dared to violate their sacred algorithm. Google's corrective action in these cases is swift and harsh and not something you want to experience. **There are no secrets to Search Engine Optimization.** Those that represent that they have one are lying to you. Search Engine Optimization is simply challenging work, deep thought, solid strategies, and high-quality content generation related to your business.

Search Engine Optimization is an important part of any marketing plan where the Internet matters to your sales. SEO is where businesses, language, people, and systems collide, and it is messy. However, it is also an opportunity that no business can afford to overlook. Start with a pace that you can keep up forever, and you will eventually win the optimization battle.

What's Better PPC or SEO?

The simple answer here is that they are not better or worse; they are simply different. Many people look at SEO, and they think it is fast, free, and easy. The reality is that it is none of those things. At the most basic level, PPC is paid, and SEO is earned traffic.

Content is the way that SEO is won or lost. Creating great content takes labor, and labor is money. What is different about this is that SEO is a one-time investment with a lifespan. PPC, on the other hand, stops when you quit paying.

SEO Lifespan

Pages can live a long time, but they do age and sometimes poorly. Google has a factor called Freshness, and it changes the way Google ranks a page. In most cases, as the page ages, it declines in this variable, and all other items being equal, this will result in a slow but steady decline in the page position within the SERP. This factor is weighted relative to the competitive pages, so if the competition rests on its laurels, you can rest on yours. If the competitor has an ongoing stream of content, you will eventually lose this battle.

Want to know how you are doing in SEO?

Run Google Ads. It might seem like a funny thing to say, but you get a quality score when running Google Ads. This quality score is a close cousin to the SEO score, and PPC and SEO share most attributes. The exception being the bid. We have several accounts that run a small budget in Google Ads for the data value of the Quality Score.

SEO is to PPC as PR is to Ads

There is no guaranteed placement in PR or SEO, but the value is much greater than the same space in paid space when you get it. The PR world likes to claim that PR is 10x the value of equal ad space. We tend to agree with them in print, but less in digital. In search, people go through phases, and in the initial stages, they do research. In research mode, users respond almost exclusively to SEO, Organic, or Natural listings (these are different names for the same thing). If they shift to purchase mode, they start to interact more with the Ads.

Use PPC to Value SEO

We use the market value established by PPC to value our investment return in SEO. If you are willing to buy a specific word in Google Ads, then it stands to reason that you accept that valuation. Use the cost per click from Google Ads and multiply that by the clicks from SEO, and you have the return on your SEO efforts. We have often done this analysis and found that SEO was often more expensive than PPC for the same click.

SEO is Sloppy

In Google Ads, we use tools to give focus to our traffic. Common in this are negative words, but SEO has no such control. This lack of control is one of many reasons that your PPC traffic quality should be better than your Organic (SEO). If it is not, then it is time to start digging into the details.

Search Engine Optimization

The Google goal is to produce the best SERP (Search Engine Results Page) for the query submitted by the user. They take this goal very seriously, and they do not like it when someone messes with that goal. Not only will they get mad about it they are judge, jury, and executioner in the internet world. Their judgments are harsh, swift, and without appeal. Major companies have crossed these lines, and Google has dealt them harsh blows. One well-known case was BMW, which is an "A" list advertiser. Google pulled them out of the SERP entirely for breaking the rules. There was a case where Google imposed a penalty on a department of Google. Don't for a second think; they will not act to protect the integrity of the SERP because they will, and it will be swift and harsh.

SEO is a balance between content for the user and the system. A page that is purely optimized for the search is horrible content. The result of not balancing this is a flood of traffic that thinks your business is an idiot. We have seen clients spend tons of money on optimization that was an embarrassment to the business.

The mission of SEO is to provide the best possible content for the SERP. Some businesses want to figure out how to beat Google at the game that Google invented. My advice to those SMBs is, *"Only try to trick Google if your programming staff is smarter than theirs."* We have

yet to meet this business, but who knows, maybe that business is out there. If you are that person call us because we would like to hire you.

First, let us give credit to the authors of this table. Search Engine Land created the table, and the infographic was designed by Column Five Media. They permit the use of this, providing we give the authors credit for their work. **So here it is – Great Job!**

SEO is not pure science. While we agree with most of these as SEO basics, they did skip over a few big ones. We believe that the page's click-through rate indicates a fulfilled search and a huge part of the SEO score. We believe this because Google told us this in this video: (https://www.youtube.com/watch?v=zStNy_PImyQ).

If you advance to 2:41 or so on the video, you will hear Hal state that the CTR is the most important, and graphically he shows this as about 65% of the quality score. Since we know that the quality score and SEO score are largely the same, we estimate that the page CTR over time is up to 65% of your quality score. Google has stated that CTR is not part of SEO, but they are always careful about how they say this. CTR by itself is not an SEO factor, but in combination with other things like the assessment of a satisfied search, it certainly is. Just clicking on a link will get filtered as it does in Google Ads.

This chicken-and-egg problem because you must have a good CTR to get a good score, but 65% of the score comes from the CTR. Without some special handling, added items would never get a chance, which is where freshness comes into play. This attribute that fades over time helps push up the latest items. There is a valid argument that the CTR is in the Quality attribute, but they do not mention this when you read the details.

For a deeper exploration of this chart, check out the SEO Manual on Search Engine Land.

After all, it is a String

Important things to consider here are that SEO factors work together and act like two ends of the same string. Businesses constantly want to pull on one end of the attribute without the other end moving, but that is not how strings work. The rule is simple; **The more things**

the page is about, the less it is about any of those things. Many SEOs struggle with this because everyone wants to rank highly for lots of different keywords. The classic plumber example is a plumber in Los Angeles who wants to rank well for Beverly Hills (a suburb in LA). The more they optimize for Beverly Hills, the lower the ranking they get for LA, and the reverse is also true. The challenge is that almost every metro area plumber wants way more territory than they deserve. The plumber in LA will want to rank first in all 84 cities that make up LA, and they do not deserve that level of ranking, and Google knows this.

The Problem with Content

Great content is hard! It takes creativity, innovation, and persuasive communication skills. Many people think they can do content, but the reality is that most cannot. They can write, but they are not professional writers. The secret to winning the SEO battle is continually generating fresh, high-quality, innovative content related to your business and marketing messaging. Do this over an extended period, and you will eventually win the SEO battle. The challenge is that generating this content is incredibly difficult because you simply run out of material after a few articles. Therefore, you must have a balance between SEO and Google Ads to win the overall game.

Solve this Marketing Equation

Explode your sales potential.

$$P(y) = a \cap m \cap t$$

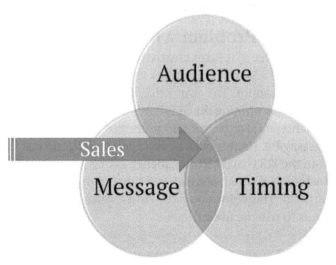

The "Yes," aka the sale, happens at the intersection of Audience, Message, and Timing. Say the right thing, to the right person, at the right time, and sales happen.

So that's the secret in 50 words.

The rest of this chapter is a deep dive into the details of how to influence these factors. Your goal is to make the overlapping intersection larger while consuming the least resources. If you want to keep this light reading, skip the next five pages because we are headed deep into the weeds.

Businesses have two resources they can use to move these circles. They are **creativity or cash**. Solve this equation for your business, and the Internet becomes infinitely scalable, and you become rich beyond your wildest dreams. The devil is in the details, and the rest of this chapter discusses those details. Take it from me; if this were easy, everyone would do it. While we doubt anyone will ever fully solve this equation, all you must do is improve the performance, starting with understanding the variables.

y = Yes

I will be Mr. Obvious here and state that yes, responses in Sales make the world wide web go round and round. The key is how to drive this number, and that is our subject for today. The first thing to consider is how much to ask for. All businesses want to jump to the final order, but this is simply unrealistic in many cases. It is a rare business relationship that goes hello to thank you in one touch. So be careful what you ask for, or you will go from hello to hell no.

Ultimately the business goal is the final sale, but you need to guide the customer through that path. While studies vary, everyone will tell you that one touch to a sale is not something that happens very often.

a = Audience AKA Your Who

Understanding and defining your Audience is the start of this process, and there are beginner's mistakes you need to avoid. First is, rookies make the mistake of thinking **EVERYONE** is a prospective customer. The problem is that 16-year-old Emily is not interested in the same things as 35-year-old Ryan. Failure to define your audience is how you create a marketing disaster. If you cannot define your audience, you are not ready for this chapter.

Audience selection is a classic quality versus quantity problem. The tighter the targeting, the smaller the audience, and if done right, the higher the quality. The game here is to tighten the target the best balance between the highest quantity and quality. Rookie mistakes are thinking that everyone is a potential customer or over-tightening to a point where you cannot reach your volume goals for the business. Marketing is a messy business, so there are no right answers. Some answers are **more right than others.**

In the world of digital advertising, there are lots of controls over whom you target, and like all marketing data, it is generally correct but never accurate.

m = Message

Think about your marketing message strategy as the painting of an image in the minds of your audience. Each touch is one more brush stroke towards completing the work of art. As touches accumulate, your marketing image emerges and forms in your Prospect's mind. Like art, different people appreciate your marketing message in different ways, influenced by their life experiences. Like art, your message needs to reflect your business why, mission, and goals uniquely. If you copy someone else, people will consider it printing, not art, and they will value you less.

Marketing messages can be both positive and negative. Current politics is a terrific way to illustrate this. Merely mentioning ex-President Trump causes a reaction: good and bad. An ad supporting Trump will get you a positive from some and a negative visceral reaction from others. Pick your side, but we can assure you that different people will take the message in different ways. Not all messages convey an easy-to-see positive and negative value, but almost every message has it. The tighter you define

your audience, the better you can fit your message and create the best communication experience.

Many will tell you to measure message delivery with reach and frequency, but this is far too simplistic. As a business, you must consider the strength and delivery cost. Reach is the volume of your target audience that sees the message. Frequency is how often they see it, and this decays over time. Strength is the degree of emotional reaction you get from the message and audience. Delivery cost is a constraint because nobody has an unlimited budget to work with.

t = Timing

Many think that markets are driven by **Need,** but we propose to you that it is **Want**. You think they need it, but it only happens when the customer wants the product or service. The transformation of a need to want is event-driven. In many cases, this is outside your control but not outside your consideration.

Let me illustrate the transformation with a true story. Years ago, our agency was contacted by a business that wanted to increase their sales. They had good service, strong execution, and a solid business that served a customer's need. We created a marketing strategy with solid campaigns with the right keywords, good ads, and bids tuned to get their site on top of the results. We turned this on, and nothing happened, and as you might expect, the client was upset. The poor performance went on for some time before we got the dreaded **"What have you done for me lately?"** call. We managed to survive this call, but we knew that our client relationship had a clock on it. We tried everything, but nothing would move the needle because the customer's need had yet to morph into a want. Then we got lucky, and the client's phones lit up to the point where they called and said: "Turn it off – We are buried!"- our favorite type of call. To create some suspense, we withheld the service involved. The client's service was snow removal, and a big snowstorm rolled into town. In this case, the poor response was known and understood, but the market's conditioning was part of the excellent ROI they saw on this campaign.

In most cases, the reason is not this simple or easy to understand, but the same dynamic is always there. Advertising can raise awareness, but it cannot make it snow.

Timing is driven by events that cause the message you have painted in the audience's mind to morph the need into a want. So, ask yourself, "what events cause the metamorphosis of your customer's want into a need?" Then look for ways to refresh your touch with them at that time. If you have done this right, they will come to you. You will never get the juggling of time and resources perfect, but you can get past the random chance you probably operate with today.

Getting it Right

When you look at the formula visualization, it is easy to see why prospecting efforts have such horrible response levels. Fortunately, your competitor faces the same challenges, so remember that you do not have to be perfect – simply better than them! Grow, shrink, and refine your audience. Fill in the blanks in your message and try to get lucky on the timing (clue: the harder you work, the luckier you get). Balance your resources and spend your creativity before your cash. We can tell you with 100% certainty which one you will run out of first – it is cash.

On-Page Optimization

On-Page optimization is simply those things you can change directly from the content of the page. SEO breaks into three large categories:

- On-Page   What you say about yourself
- Off-Page   What others say about you
- Votes   How it performs in the SERP

> **"Give us the skills to change what we can, the knowledge to know what we cannot, and the wisdom to know the difference."**

This chapter is a dive into some of the details of how on-page optimization works. The simple definition of on-page and off-page comes from the source of the data. On-Page means you can directly affect the score by changing the page. Most Off-Page items cannot be changed directly. Nobody outside of Google knows what the weight of on versus off-page is to the total score, but my guess is more than 80% comes from off-page items. Another way to think about this is that on-page is what you say about yourself, and off-page is what others say about you. Every professional we know realizes that what others say about you is the core of your reputation, and this is true both online and off.

"Google gives us the skills to change what we can, the knowledge to know what we cannot, and hopefully we have the wisdom to know the difference."

The first on-page concept you must understand is the frequency sweet spot. The most common keyword target, after your business name, is typically your service or product and the geographic location. Using our Plumber example, this is "Los Angeles Plumber" or "Plumber Los Angeles." Density is how often the keyword exists relative to the total content. If you have three uses of the word plumber out of 100 words on the page, you have a 3% density. The initial reaction from most people is to stuff more keywords into the text, but this is a sweet spot, so more is not **always** better. For a sweet spot, more is better, up to a certain level, then more becomes negative. Go way over the line, and it will turn into a penalty called Keyword Stuffing. The best guidance we can give is to give Google what they seek. The best quality content and they greatly value balanced editorial copy. If your content reads like an editorial article, it is probably in the right range. Editorial content has a balance of positive and negative terms compared to sales content that is almost pure positive words.

Keyword Stuffing – Don't do it!

> ## The More You Are About One Word
> ## The Less You Are About Another

Overstuffing keywords is a common rookie mistake. We have seen way too many overstuffed pages, and honestly, they are an embarrassment to the business. These pages read like they were written by a three-year-old with a crayon. You must ask yourself why you would want to attract traffic to your site just to embarrass your business with a horrible presentation.

For some reason, this is difficult for many to understand, but **the more you are about Plumbing, the less you are about HVAC or**

Remodeling. Plumbers often do not understand this, and we can always tell when this will be an issue. We meet with the client, and they produce a list of keywords normally with 25-50 great keywords. They hand me the list and say, "I would like to be first on Google for all of these." We hand the list back and ask them which one they want. Small business only gets a few core words beyond their business name, and they always want more than they deserve. **There is no way** a small business is going to rank first for 25 to 50 competitive keywords. The simple truth is you are just not that important. So, you must pick your target wisely. We will give you a clue on this one – if you are a plumber, then the keyword you want is Plumber, followed by the major city in your market. No matter what type of business, I can assure you that you already know your prime keywords.

Support Your Keyword

Now that you know your keyword, you want to make sure that your website targets that word. If the keyword is "Los Angeles Plumber," then this phrase should be in the:

➢ Page Title
➢ Meta Tag Description
➢ H1 Header (and there should be only one of these)
➢ Body Text

Page Title

Page title shows in the top bar of the browser, and most people never notice it. Technically it exists between the <title> and </title> HTML tags. It is also commonly the headline of your organic listing. Here is something for you to think about, the title has a lot to do with your click-through rate on your listing because it is just like the headline of an ad. An Advertiser that runs an ad with their business name as the headline usually sees an extremely poor response unless the search was specifically for your brand. The best headlines connect to the search query. If your keyword is **Los Angeles Plumber,** then the most responsive headline is **"Los Angeles Plumber**." Remember that there is a limit of 55-60 characters for a title. More than that, and Google will just cut it off. If your business name is Joe's Rooters & Pipes, and you put that in the title, guess what reaction you will get?

Meta Tag Description

The Meta tag description is the description of the page, and your keyword should exist within it. Think about this as the body of an ad because that is what it is. One common use of this tag is to create the snippet under the link on the SERP, and people do read this! It may not be visible on the page, but it is visible in the search engine. Take the time to write a great description of what is on the page and keep it short. If you go on and on, Google will just cut it off at the most inopportune point. Do NOT stuff keywords here because if the rest of the page does not match the intent in the meta tag, then Google will ignore it or worse. Within the HTML, you can go on forever, but Google will only use 130-160 characters.

H1 Header

Web pages have a hierarchy designed into them, and some designers think it is better to use CSS and create their styles outside of this hierarchy. This design decision forces the search engine to figure out the new hierarchy, forcing Google to guess. Use the simple H1-6 tags but feel free to take advantage of CSS and override the default values to get the look and feel you want. You should have one H1 tag on a page, and it should be the most dominant word on the page. It should also be your primary keyword. Sometimes people want to do something cute or humorous on the page, and that is fine. Just do it in graphics and keep your keyword in the text. You can reduce the size and position of the H1 tag to make your presentation work.

Body Text

Body text is all the other text on the page. It has the least amount of weight in the calculation, but it needs to continue the support of the keyword. Body text will have a slightly different value for bold, links, lists, and other attributes, but it is all body text.

Keyword Order Counts

Most keywords are phrases and contain more than one word, the order of the words count. The keyword "Los Angeles Plumber" and "Plumber Los Angeles" are different keywords. Google Ads will override the word order and match both, but there is a difference in the finer art of optimization, so you need to pick one. Dominating one will not exclude you from the other in this case, but again there is a difference.

When it comes to keywords, the more words joined like this, the more specific the target and the better the optimization.

Don't Try to Trick Google

Optimization Rule 1.01 should be that you should only try to trick Google if you think your programming staff is smarter than theirs. Occasionally, a trick will come along that works for a little while, but Google always figures it out. When they do, they come down on the violators with the wrath of Google. We have seen many that got caught trying to trick Google. The trick angered the Google gods, who banished the offending sites from the index. You may not like it, but in these cases, Google is the judge, jury, and executioner, and they are exceptionally good at the last one.

Page Speed

Page speed is part of the user experience, and being too slow can cost you lots of SEO points. Speed and experience are a balance that all websites wrestle with, and there is no perfect answer. In most cases, the rich media expected in a modern website slows the website but enriches the content. We recommend the Google Page Speed Insights tool currently at this URL.

https://developers.google.com/speed/pagespeed/insights/

Mobile Friendly

Google has been in Mobile First mode for several years, and this means that you should be concerned if your mobile speed from the tool above is below average. The experience and goals for a mobile user are different, which must be part of your design considerations.

Security

Not having a secure site is an epic failure in SEO and has been for some time, yet we still see these every day. If your website does not load with an https (https://www.yourwebsite.com), talk to your hosting company today and have them fix this.

How to tell when an optimizer is lying to you

This is easy - their lips are moving. All joking aside when someone tells you that they can optimize your page and get you first on

> # Winning words nobody searches for is easy, but not valuable.

Google - They just lied. Most of the page score comes from off-page, and while you can do things wrong on the page, it is **impossible** to do something on-page that dominates a search result. There are tricks involved in this process, and one is the **illusion of results** based on freshness. One of Google's attributes is how fresh the content is, so the straightforward process of making changes will move a page up in ranking, but the change is temporary at best. Another game played here is optimizing for words nobody searches for. It is amazingly easy to optimize and win a word that nobody else wants, but the reason nobody wants it is because nobody searches for it.

My Best Optimization Advice

The best results for SEO optimization come from organizations that write non-stop about the topics relevant to their business and are passionate about producing original content. For businesses that lack the topic passion or writing skills, this is rarely done. If you want to dominate the organic traffic, you need to start a blog and commit to writing one article each week for at least six months and do so with lots of pictures and other connections. Google is searching for useful content, so if you create it, they will eventually find it. Share your knowledge with the world, and Google will reward you with great organic positions. Hold your secrets close to you, and you will need money to pay for the traffic you could have earned.

This book is far from a complete treatment of this subject, but we would like to share one more thing. Every success we have ever seen in optimization did not start with the mission of optimizing. It started with people who are passionate about their topic, wrote with that passion, and connected to the audience. Google is always trying to find great content for their search results, so write great content and do not hide it from them.

Off-Page Optimization

What others say about you is more important than what you say about yourself

Off-page optimization is simply all those things not under your direct control. The biggest SEO factor is back links, and the value goes back to the concept of page rank, which is the essence of a Google Search.

What others say about you has been the core of business reputations and referrals for generations. On the Internet, what they say is expressed with backlinks, and backlinks are the core of the SEO score.

Earning comments is key to winning the SEO battle. Let us stop for a second to think about what it takes to get someone else to say something good about your business. To do this, you have to create a **WOW moment that is worth talking about**. Wow-moments come from creating and delivering an unexpected level of service. Getting customers to talk about your business is one of the hardest things businesses must do in brand building.

Step 1: Get the easy ones out of the way

Almost every business belongs to organizations like Chambers, Industry Groups, and community groups like schools, churches, and many others. These groups maintain rosters, and you need to make sure they are accurate and use the same physical address. The address will help with map citations, and the backlinks will help with general SEO— the more of these, the better. If you are active in one of these groups and hold a position, ask for a backlink from the site.

Step 2: Get your social media up to date

Almost every business should have a Facebook and LinkedIn account, and it does not hurt if you have Twitter and Instagram. Ensure that the business location data is accurate and that you have all your employees attached to the company profile. If you are already active on social media platforms, make sure you occasionally mention and point to your site. Not so much that it becomes spam, but often enough that search engines and social media sites understand the connections.

Step 3: Now the Hard Work Starts

Now you are into the work of earning backlinks, and it's way tougher than it sounds. Here is a basic list of places to get started:

1. Search engines and Directories
2. Industry Forums
3. Media Sites (Newspapers, Radio, TV, and others)

Search Engines and Directories will typically have submission rules and portals. Many tools automate submissions but check the tool reputation before you use it. Some tools do an excellent job, and others are scams or worse. You do not want to continually submit your site over and over as some tools do. Bad tools can get you on the bad kid's list, and there will only be a lump of coal in your SEO stocking at Christmas.

Industry forums allow for conversations between different people with like interests. Here are examples for Plumbing, and almost every industry has similar hub sites.

➤ www.plumbingzone.com
➤ http://www.plumbingforums.com/forum/
➤ http://www.contractortalk.com/f9/

Search forums, and you will find lots of these. You need to watch the activity for a while to assess the actual value of the listings. If the forum appears to be inactive, it will just waste your time. There are forums for DIY (Do It Yourself), and the same rules apply except in those cases you are involved as the expert answering the questions. Again, you must assess the balance between time consumed and value created.

Media sites generally are the most valuable and the most difficult. If you have a business story of interest to their audience, then pitch it. If they write it, then promote it. An editorial listing on a media site is worth its weight in SEO gold. The key here is to decide what depth makes sense and consistently engage with the outlet.

Word Wars are not just SEO

At first glance, it might be easy to confuse Word Wars with SEO. SEO is one of the battles in a Word War and the first one people focus on. Starting with SEO can be a strategic error it takes to win SEO positions. The War is waged across multiple conflicting battles, including Search PPC, SEO, Web Experience, Social Media, Display PPC, email, and many others. The Battles are related to the War, but you must remember it is easy to **win the battle and lose the war** by lacking balance in your strategy.

No Right or Wrong Just Different

Over the year's clients have said to me, *"Tell me what to do, and we will do it. "*and that assumes that there is one right answer. **There are no right answers** – just opinions that change all the time. English is a complicated language with 171,476 to 470,000 words. Add to that industry-specific terminology, and the number quickly reaches over one million words, and most words have multiple meanings. The statement made famous by President Bill Clinton, *"It depends on what the meaning of "it" is, "* is fair because in the dictionary, the word "it" has 72 meanings. The point Clinton was making is that words have many meanings.

What you said is not what we heard

On top of the complexity of the language is the extraction of meaning driven by the reader's life experiences. If you want to win a Word War, you must understand how your audiences translate your words into meaning. All readers start with the perspective of your message, and it is from that perspective that they derive meaning.

Empathize and Serve Your Audiences

Audiences are a collection of individuals defined by shared attributes. Audiences can be created from structured information or behavior, but the result is the same it is an audience. One of the first steps in any Word War is to decide who your audiences are, and if you say everybody, then go back and think again. Having an audience of everyone is worse than having no audience at all. It is important to have a persona for your audience so you can relate to your audience personally. Write out their attributes and get a picture and give them a name – make it personal.

Word Wars are Not Just Text

In current technology, Word Wars are not always expressed purely in text. Word Wars can be audio or visual, but ultimately, they tie to words. Images contain meaning that can translate to the concept of words. In the AI field today, there are tools like Amazon Recognition (1) or Microsoft's Computer Vision (2) that create text tags for images. If not already in the major search engines, it is coming fast.

Points versus Distribution

Understanding a Word War requires an understanding of the foundational process. Your content strategy creates eligibility and distribution. Page-Rank creates points for this distribution. Page-Rank takes many people a while to wrap their minds around this because it is iterative logic. You calculate page rank once, then do it again because you changed the page rank of the sites.

Here is the original Google formula and the core of how it works today.

$$PR(A) = (1-d) + d (PR(T1)/C(T1) + \ldots + PR(Tn)/C(Tn))$$

The full discussion of this formula by the founders of Google is available at the Stanford University website.

Simply translated, the more pages that point to your page, the higher your page rank. Who is doing the pointing, and how unique the link also counts. This fundamental process in the Google Search Engine has not changed for decades, although there is now a very thick layer of protective logic to stop it from being tricked. This enforcer layer puts penalties on your site if you try the trick of the day. My advice to clients has been the same for years.

Only try to trick Google if your programming staff is smarter than theirs.

If you try to trick Google, you will get caught. They respond quickly and with absolute power. The cost of this can range from a page penalty to your entire site being removed from the Google index. The stakes are high, so be careful out there.

You build your points the same way you build a real-world professional reputation by sharing your expertise with the world. If you want to rank high in the search results, **create content worth talking about, and don't hide it** from Google. You do not need to trick them; you just have to serve them.

Content Strategy

The first step in developing a Content Strategy is to decide what your brand is about. Start by understanding your audiences and the reality that you cannot, nor do you want to win all the time. Because of an immutable marketing rule that says, *"The more things you are about, the less you are about any one of those things."* Every part of your content strategy holds the potential to make a friend or enemy, based on the message match.

Consider the brands of Walmart and Tiffany, both world-class brands and companies. Going to a Walmart audience with a Tiffany strategy is a disaster. Walmart is a low-price brand, and Tiffany is a high-priced brand. This example is a message to audience mismatch, and that never ends well. They are not better or worse. They are simply different. You must pick your battles and serve your audiences.

SEO Elections

EVERYONE MATTERS!

Google tells us that the organic CTR does not count in SEO, and after watching search engines for over 20 years, **we must disagree**. We do not think that Google is lying to us, but we do not think they are completely transparent. We believe that the CTR by itself is not used because the data is just too dirty. What we believe they use is the validated and cleaned CTR. We see them use the technology to clean this data almost every day. They scrub Google Ads data and remove the suspect click many times every day. It is the net CTR after cleaning that we propose they use.

When this competition happens, just a few things from the page play a role in the voting or CTR measurement. They are:

> ➢ Website Name
> ➢ Page Title
> ➢ Page Meta Description

smsrd.com ▾

Google Ads Expert | Get More Calls with Smart Digital Marketing

More Calls, More Contracts, More Clients - Certified **Google Ads** Agency providing complete marketing solution to grow your business.

This listing above is an example of the SERP listing. You can see that the page title is the dominant data in the listing. It has a larger and bolder font than the rest of the listing.

The Meta Description is next in importance to the listing, and you will hear this called the page snippet. It is not always the Meta Description from the page. If Google feels that the Meta Description is misleading, the algorithm will pick a snippet it believes better reflects the page contents.

At the time of the click by the user, this is all that exists in all listings. There are specialized listings that are enhanced from this basic listing, and most of those will improve the CTR of the listing.

Before someone runs off to click on their site until it is number 1, understand that Google figured that trick out decades ago. Google cleans this data daily, and if they think you are trying to trick the algorithm, it will not end well for you. Google **will take swift and harsh action on your site.**

Many people are surprised when they see their listings and realize the horrible job done on this data. If you want to see your site all at once, use a site search. A site search starts with the word site followed by a colon and then your website address. Here is an example

site:smsrd.com
Just to recap SEO, it is:

- ➤ What you say about yourself
- ➤ What others say about you
- ➤ Election results

Optimizing Maps

Maps for SMBs are high-value organic positions that exist on most searches. Unlike keyword-organic-listings, maps are eligible based on address citations, then ranked based on the reviews and keywords. The weight of a review depends on the authority associated with the review. There are only three map listings on the first page, and there are always more than three eligible SMBs. Who will earn the coveted three positions depends mainly on the reviewers' number of reviews and reviewer authority? Let us take this apart and examine how each of these works.

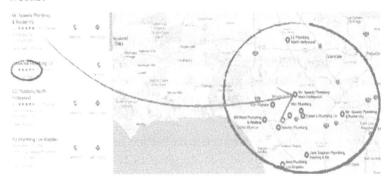

Keywords are checked to see how the site ranks for the search. For this, you can refer to the SEO section that discusses the Keyword Organic Results. While the site's authority will impact this result, you can think about this as a yes or no eligibility test. Should the site appear, and in most cases, the answer will be yes for way more than three sites.

Citations come next, and Google is looking to verify that the SMB is in the city included in the search. It will look at the website and other sites to see if the address is related to that site. The more times it finds the same address, the more it will trust the citation, but it stops counting at a certain point. Five citations from good authority sites are just as good as a hundred because once it accepts the location as valid, it passes the eligibility test. Additional citations after will not help, but they also do not hurt.

Step 1 – Claim Your Listing

To start, you need to claim your listing and get the information on it accurate. This step only gets you into the game. The hard part is getting your site on the highly coveted top first-page listing, which means citations and reviews. You do not want to promote a site that you have not claimed and never let someone outside your business own the listing.

Citations

Citations are other sites that point to your site that include your physical address. The best examples of this would be your listing on the Better Business Bureau or your Local Chamber of Commerce. While these are the common ones, they are far from the only ones you should be concerned with. If you belong to a trade group that links to its members, make sure your data is up to date and includes your address. Almost every community has directory sites, and you want to make sure your business is in these, and they have the correct physical address. If you carry specific brands, make sure you are on the manufacturers' dealer locator as these **are high-value citations**.

Citations: Trust & Authority

Not all citations are created equal, and the citation's value is relative to the trust, authority, and relevance of the site that is pointing to yours. The higher the website's profile, the more juice it gives to the citation, which is the same as the SEO process of simple backlinks. Google is surprisingly good at knowing what sites to trust, and you will find that these are the same places people trust offline most of the time. The BBB and Chambers are examples of this because they are worth much more. After all, they are high trust sites.

The Most Important Citation

Your website is the first, simplest, and most important citation. We have reviewed thousands of local service SMB websites, and it amazes us the number of times we find the website missing the physical address and phone number. The site should be connected to the Google My Business listing since that is the first place Google would look for the support of the map listing. Your address should be in text format on the bottom of every page, along with your phone number. We realize that nobody shows up at most in-home service SMBs, except for the few with showrooms, but finding the address is key to getting on the map.

Reviews

Reviews count, and you need to ask for them. When you have completed a job, and the customer is happy with the results, give them your card and ask them to provide a review of your work. On your business card, make sure you include instructions on how to enter a review. There are thousands of places on the web to leave reviews, but we recommend that you focus on just a few; Google and Yelp. Reviews are powerful, but power can swing both ways. When they go negative, you need to be ready to deal with this. The first rule is to keep your cool and respond to the negative review. If you know who posted the negative review, **call them,** and resolve the problem. If you do not know the person making the negative post, respond with a public offer to make it right. Never respond in anger and never get into a fight on a review site. There are times when the best business advice is just to turn the other cheek.

Conflicts

Businesses want to appear in cities that they do not have physical offices in, which is a great challenge. If your business serves more than one city, then simple logic says you want to appear in all the cities. The problem with this is that these positions are competitive, and competitors in those cities get the available slots first. You can, and should, define your service area in Google, but that will not get you into these highly competitive slots. The problem here is that the more you are about one place, the less you are about others, so you must decide what your primary city is and compete for that. The odds of winning a second city are somewhere between zero and null.

Gaming the System

The "Trick of the day" might create a win for a brief period, but Google is brutal when they find someone gaming the system. The normal response from Google on this is to rip the entire site out of the index or put it into penalty status. If you think that they will not do this to you – **you are mistaken**. We can guarantee you that you will get caught at this. Google will call and mail to the location to validate it, or your competitor will complain, creating an investigation by Google.

Summary

Your business must be on the first page of Google, which means competing in all the sections. For local service SMBs, this includes Guaranteed Service Ads, Google Ads, Maps, Images, Organic, and whatever Google invents next.

Track Everything!

We are huge fans of tracking early and often. Some businesses think they should only track orders, but that is a huge mistake. Most businesses have a long sales pipeline, and you need to track each stage in the sales funnel. The exception is emergency traffic, which commonly goes from click to call to visit in one transaction. Of course, you want to track that high-value conversion but watching just that is a dangerous strategy.

Tracking can get extremely complicated amazingly fast. As much as you might wish that the path from the first contact to the final deal was a straight line, it is NOT. Therefore, you want to track this as early as possible and everywhere in between. We are going to use a Bathroom Remodel as my example in this. In the preliminary stages, the customer is researching their options. In this phase, they are attracted to content that answers their questions. As they get further into the purchase cycle, customers start to pick businesses for their project shortlist. Then working down from the shortlist, they select the service providers. This entire timeline can easily take 90 days and involve many visits to your website, phone calls, sales appointments, proposals, negotiations, agreement, execution of the work, and payment.

Website Measurements

If you have a website, you need to design measurement into it. Without measurement, your advertising investments are entirely faith-based. Leaving you making decisions based on; **what you think you know - not what you know.**

Site Analytics

We always recommend Google Analytics for site analytics because it is free and integrates tightly with Google Ads. You do not have to have Google Ads running for this to work, and you can get a free Google Analytics account. Running a website without Analytics is a marketing crime of the first order.

Google Analytics is a script-based analytics product, so the script must be installed on **every page**. If the code must be installed by hand on each page, ensure that you validate every page after installation. Most CMS (Content Management System) packages will have a template for the header or footer that goes on every page where you want to install Analytics. The typical recommendation will be for the footer, so it loads after the page is visible to the user. To validate the installation, simply view every page on your site and then check the next day to see that every page has some activity. Google Analytics is a complex product, and there are exceptionally good books on this topic. If you want a deep understanding of this topic, we recommend the book Advanced Web Metrics by Brian Clifton. We have read both versions of his book, and they have helpful tips and pointers on understanding your traffic. The bottom line here is - **Install Google Analytics now!** Operating a website without Google Analytics is a high crime for a website operator.

Form Responses

Forms are a popular way to allow visitors to contact you, and we highly recommend having this available. Once data is entered into your website's form, the information is then emailed to you, and the visitor is routed to a thank you page. Web Designers like to avoid the thank you page, but you should insist on a separate thank-you page for tracking purposes. One major advantage of a form over a simple email link is that you can more accurately track your conversions. Another bonus is that you can control the form's fields, ensuring the visitor answers the essential questions you need to help them. There is nothing worse than getting a great sales lead without a way to contact the customer. The guidelines for the form are that you should **ask for everything you need but nothing more** than that. When the form asks for and requires more information than you need, it can lower the response rate. Every time you ask for more information on a form, the person may decide to leave. Web traffic is skittish, and visitors can be easily spooked. Be careful about how much you ask on the first contact.

If you use an online form on your site, a spammer will find it. Spammers will try to use your form to push email at you. The way to slow this down significantly is by using a CAPTCHA. CAPTCHA is an acronym for **C**ompletely **A**utomated **P**ublic **T**uring test to tell **C**omputers and **H**umans **A**part. A CAPTCHA is a challenge-response test used to decide whether the user is human or machine.

You can get public domain versions of this for free, or you can install simple logic tests into your form. If the problem is bad, then the more sophisticated CAPTCHA makes sense but, in most cases, simply asking them to key a specific value is enough to make most spammers go away. To give you an idea of how prolific these spammers can be, one of our guys in the office has a blog with almost 5000 blocked spam comments from the last two weeks.

Phone Numbers

There is no doubt that good websites **make phones ring**. Connecting phone calls to your web analytics is a challenge, but it is possible. You can call your phone company and get what is called a tracking phone number and point it to your primary phone line. Using the tracking number only on your website, you can separate your web phone calls from all the other forms of advertising.

The next step is a little more complex, but it gets closer to the real advertising contribution. In this implementation, you get two tracking numbers. The website displays one if the traffic is from advertising and the other when it is not. This process is not perfect, and when it fails, it tends to underreport advertising contributions, but it is better than having all the calls lumped into one big bucket.

Do YOU own YOUR phone numbers?

In some phone tracking or advertising service agreements, the service owns the phone number. You cannot let that be. **Make sure that you own any phone number** connected to your business. We've seen agreements where the service or ad network owned the phone number and held the business hostage when they wanted to make a change. If they wanted to keep getting calls from the numbers, they had to renew.

Customers routinely save phone numbers on their phones, and lots of business calls come from that.

Email Contacts

Some businesses are not crazy about email forms and prefer to put their full email out on their site for people that want to contact them that way. Technically this is done with an href tag and a mailto: in the link. The problem with this setup is that you cannot track that event without special coding. However, if you have Google Analytics installed, you can extend the email link with an "on click" event to track the email interaction.

This page will have a visit logged every time someone opens the email link. It cannot measure if they sent the email, but you will know if they started to send one. You can also link this page to a goal in analytics and then feed that back to Google Ads.

Conversions & Goals

Conversions and Goals are ultimately what most people want to measure. Conversions are from Google Ads, and Goals are from Google Analytics. Goals include actions from all sources of traffic, and Conversions are ONLY from Google Ads traffic. We recommend a separate thank you page because that gives you some place to install the conversion code and gives you a separate page for the goal tracking. With Goals, we recommend that you set up the full URL path that takes them from the landing page to the thank you page so you can see the loss at each step in the process.

This book is far from comprehensive coverage of tracking responses, but if you do at least this much, you will get some great data that can help you understand how your advertising returns value to your business. What is a Good Response Rate? Response rates vary based on many variables, but in general, a good response rate for a Plumber is in the range of 25-30% when you count both online forms and phone responses. The bulk of emergency service responses, no surprise here, are phone calls and online forms are just a backup for after-hour inquiries. The truth here is that results vary based on lots of factors. There is a significant difference between emergency and routine services, and the keyword can often give you a clue of what type you are dealing with.

Emergency traffic will typically stop shopping when they become convinced that the business will take care of their problem. If your website or phone staff fail to meet that level, customers will keep shopping, and trust me; a **voice mail message does not meet this level.** Customers have zero faith that leaving a message will get a response. They want to hear from a person that something is being done now to solve their emergency. For emergency traffic, every position lower than first on the screen is a risk that the person will stop seeking your service - position counts! If you are in the emergency services business, ensure that you have an excellent after-hours service that does not act like an answering service.

Routine traffic will search much deeper into the results, and they will talk to or review 3-5 businesses or more before deciding whom to contact. With this traffic, you need to be on the shortlist, but you do not have to be first. If the person shops five deep, then being first simply drives your cost up.

Nobody is ever happy with their response rate, and you must continually work to improve this. Here are a few typical response rates that we have seen over the years.

Type of Traffic	Response Rate
Emergency	20 to 30%
E-commerce	2% to 5%
Engaged Reader	3% to 7%
Subscriber (No Cost)	1% to 2%
Free Offer	2% to 7%
Leads	.5% to 3%
Phone Calls	.5% to 2%

The general rule of Thumb for traffic is that the higher the commitment, the lower the response, and the more you ask for, the less you get.

Agree to Disagree

Digital Marketing is often more art than science, and reasonable professionals can disagree when looking at the same data set. Based on this, it should be no surprise that we do not agree with every "Best Practice" published by Google. We recently ran into one of these conflicts, and we thought we would share it.

Google has a best practice that includes having three or more ads for each ad group, and they claim that it improves CTR by 5%. This claim is followed by another best practice that ad groups should be set to optimize by clicks. The claim here is that this improves CTR by 5%, and we accept that this is what they believe and that there is no attempt to deceive anyone. It should be noted that I would propose that both are the same 5% because you need both settings to get the effect. I do not question their data, only their conclusion. As you might have guessed, we do not agree with this "Best Practice," and here is why.

Whose wallet are we concerned with?

CTR is a measure of performance important to Google because **that is how they make their money.** Google makes impressions and sells clicks. The effectiveness of that conversion is expressed as the CTR. The challenge is that for my clients, CTR is not the key metric. Their metric is always some form of conversion where business **value is created for them**. CTR is a measurement on the path to value creation, but it is not creating value itself. This fundamental difference is what causes us to look at the data from a different value perspective.

Let us look at the multiple ad copy recommendations first. They get the result with multiple ad copies because they control the rotation method, and as they say in Vegas – the house always wins. The odds of finding an ad that increases the CTR increase with the number of ads available for testing. The downside to this is that it destroys the AB test that advertisers should be running to find the best copy. The Google algorithm is amazingly fast to pick a winner, much faster than any professional recommendation we have seen on split testing. There is no doubt that this is a winning process for Google at a Big Data level, but for your small business, that might not be. We believe that our goal should be the best copy, not just the best CTR. Good ad copy is designed to get clicks from the right people, not just the most people. Remember, most people – good for Google; right people – good for advertisers.

History Lesson

Rotation is an area where Google and Advertisers have had conflicts in the past. There was a time when Google went beyond recommending rotation and removed the option forcing advertisers to do it their way. The push back from advertisers was so strong that Google backed off and put the option back and made it a best practice not to use it. Not their finest moment, but it shows how valuable Google thinks this option is. We commonly run this with forced rotation because we are running split tests, and we want to decide who the winner is. Google calls a winner much faster than we would recommend. We normally want to see a confidence level over 80% on an AB test before deciding which ad is the winner.

Google Believes

Google believes that CTR is the way to measure performance, but they will do conversions if you have enough data. The problem is that they are not considering all the other variables that are important to your data. We propose that Google Ads are a conduit for demand and not a demand generator, so the results are subject to outside influences. Therefore, it takes time to test. It is demand in the searcher's mind that starts the process, and nothing in Google Ads influences that. The best example of this was the Snow Removal Client, who tried everything to get his conversions flowing, but nothing worked until a huge snowstorm hit. Trust me when we tell you there is no "Start A Snowstorm" feature in Google Ads.

Testing

Testing is a complex process, and it is never as easy as a single number. Many variables impact test results, and we have seen many conflicting test results, where "A" won the first test but lost the second. We often run margin of error tests with the same ads simultaneously and get different results. In the margin of error tests, we have found on CTR that a range of plus or minus 10% is common, so when someone tells me that something is 5% better, we want to see how big the dataset is. If your normal CTR is 5%, then the real CTR is somewhere between 4.5%-5.5%, and anything in that range is too close to call.

Does this mean that the Google approach is wrong – NO! It just means that there can be different right answers. The Google method has some strengths, and for some accounts, it could be the best answer. The key here is to understand the strengths and weaknesses of each option and then apply the process that is the best fit for your specific situation. In an extremely broad sense, Google Best Practices work best for large high-volume accounts but are not always a great fit for small businesses with highly constrained budgets.

Smart professionals with good intent and the same data can come to vastly different conclusions. These complexities are what make marketing such an exciting occupation.

Google Analytics

Google Analytics is a free tool from Google that integrates with Google Ads and tracks the performance of your traffic after they arrive at your site. To operate a website without Analytics is a crime against your wallet.

The first step to implement Google Analytics is adding the tracking code to every page on your site. This small script is commonly placed on the standard footer, so each page's inclusion is taken care of in one update. The best practice today is to use Google Tag Manager, and that effectively does the same thing. You can download the script by signing into your Analytics account then click on the Admin Tab.

Under that, you will find a tab for .js Tracking Info and under that Tracking Code. This process gives you the code that you need to copy to every page on your website. When this code executes, Google Analytics will register the page traffic. Analytics is a script-based tracking system, meaning that this code must run to measure the page. The other type of tracking is server logs that pick up all the activity regardless of the script. The difference between these two approaches can create debate because the numbers will not always match. They do not match because they use different recording methods. Scripts do not pick up robot traffic, but server logs do, just one of many differences.

All you do is copy the code in the box and paste it into the page footer. Technically the code can go anywhere on the page, but we recommend the footer, so you only register full page loads. Putting it in the header will make the count go up, but the added visits are of no value because the page never finished displaying. The most common cause is a quick hit of the back button when a page does not load fast enough for them. Once the pages have the code, Analytics will start tracking visitors and reporting them within the system, and that is when Analytics gets fun.

Did they get there?

For an SMB, there are a few basic items that everyone should do. The first is to ensure that the traffic purchased gets to the site. Traffic can be expensive, and making sure you get what you paid for is a really good idea. Many might think that the clicks you paid for in Google Ads and the visitors you get are the same things, but they could not be more wrong. While you might think these things are perfectly connected, there are hundreds of steps between those two events. The big one is the time it takes for your website to respond to a click request, and the longer that amount of time, the more likely the visitor is to stop the page load, typically by using the back button or clicking on another link. At this point, the damage has been done, and your account has been charged. We do not want to tell you how many times we have found nonresponding websites with expensive paid traffic pointed at them. In our practice, we have a system that checks websites every hour and instantly reports any outage to our staff for corrective action. In many cases, the site will fix itself, but sometimes these outages can go on for hours.

How long did they hang around?

For an SMB, the time-on-site measurement is a big deal because the longer a visitor hangs out on your website, the more likely they will call you. This early indicator of traffic quality can measure the quality of traffic before other things like orders or phone calls start to happen. Analytics allows us to measure the time on site and trigger goals from that. Google Ads allows us to import goals from Analytics, so it is possible to know what contributes to the phone calls.

How is organic traffic contributing?

There is little or no doubt that organic traffic contributes to website success. The challenge is that it is exceedingly difficult to get facts to prove this. Over the years, Google has reduced the amount of information they give to websites about organic traffic, which is truer today than in the past. It is common to get less than 10% of the keyword data for organic traffic, so only in large volume sites do we see enough traffic to estimate this accurately. We can tell you that organic traffic breaks down into generic and branded traffic, and mixing these makes understanding your site more difficult. Generic traffic comes from keywords related to the subject of your website but not your brand. Branded traffic is from searches specifically looking for your business. Branded great quality, but there is never enough of it.

What did they read?

What a visitor reads is an important question for some types of traffic. For example, if Water Heaters are an important part of your business, then traffic from the home page to your Water Heater page is important, and arriving at that page creates value. For this sort of thing, you might want to set up a goal in analytics that you import back to Google Ads to show what campaigns, ad groups, keywords, and ads contribute to this value creation. It is important to track the number of pages loaded to the number of actual jobs performed, so you start to learn the relationship between these two data sets over time.

What gets the credit?

Let us start with a hypothetical job ticket. You are the owner of Joe's Plumbing in Los Angeles. The consumer goes to Google and searches for "Plumber Los Angeles," clicks on an ad, and lands on your home page. He writes your name and phone number down. Then continues shopping and visits 5-6 other local plumbers. That night the husband talks to the wife about replacing the water heater, which they both know is failing. They discuss the different plumbers trying to figure out whom to call. The wife mentions that she sees a sign for Joe's Plumbing (that's you) at the soccer field where little Timmy practices each day. The next day, the husband searches for Joe's Plumbing Los Angeles, and your site comes up organically. He clicks through on that, looks at the water heater page again, picks up the phone, and asks for an estimate. Two hours later, he calls back and schedules the job. You ask the customer how they heard about you, and he says he found you in Google under your name.

The example above is not an atypical story, and we have seen cases that are even more elaborate. With these facts, what gets credit for the order? Tough question, and the closest to correct answer, are that all the touches should get credit. The challenge is that it becomes too difficult to track. Organic could get credit because of the search for your name. The sponsorship for the soccer team could get credit since it was one of the factors in making the call. Google Ads could get credit for getting the process started. All of these are valid arguments for crediting the revenue to specific marketing investments. If you take out any of these links in the story, the result could be a business that never happened.

Social Media

Social Media is a broad term that describes a class of websites designed to connect people. Over the years, the big sites have changed and could change again in the future. Today, Facebook, LinkedIn, and Yelp are the three dominant social media networks, but there are many others.

Social Media has changed the way communities communicate, which directly affects the referral process within your marketing. For generations, home services businesses have succeeded or failed based on "What others say about them." With Social Media, this voice has been amplified to a whole new level.

If you think only in terms of emergency service, then social media is a minor factor. However, you need to look further out than your next phone call. Social media is where people react positively or negatively to your business, and you must be aware. In the olden days, a bad customer experience would get shared with 12-20 others. In today's world, that number is MUCH higher, with hundreds being exposed in milliseconds of a social media posting. What others say about you travels at warp speed in the new social media world.

Social media must be part of your overall marketing plan, and for local service SMBs, some advertising dollars need to go here. The biggest mistake businesses make in social media is that they try to control the conversation. You can engage with the conversation, but you have no control other than over yourself.

We will discuss the primary social media networks you need to be aware of on the following pages. Social media is a rapidly changing area, so make sure you stay up to date.

CEO Guide to Social Media

All my life, my friends have been CEOs of SMBs. We have been a Digital Marketing advisor to hundreds of businesses over the last 24 years. Since their inception, we have been involved with social media networks starting with MySpace and creating Kids Around the World around 1995. This chapter is an executive overview of how to keep control over this area within the business.

CEOs hold a unique position within their organization. They report to nobody but are responsible to everyone. My friends sometimes joke that the CEO has more bosses than anyone. To some degree, this is true, depending on your perspective. What we want to explore in this chapter is a high-level strategy that we propose fits most CEOs. This tactic fits any organization where the CEO has a key market-facing role. In this chapter, the CEO means the highest-ranking marketing-facing executive in the business.

This guide is NOT for communicating from the Executive to the Consumer Market. Social Media is a tool that can be used in many ways and for many purposes. The scope in this chapter is communication to a Professional Network, NOT a Consumer Network like Facebook.

CEOs have the greatest communication challenge of any executive because, in almost all cases, the other executives have a subset of the responsibility of the CEO. In a typical business, CEOs need to be aware of and sometimes communicate with Clients, Customers, Prospects, Vendors, Employees, Partners, Competitors, and more.

Rule 1: Pick the Right Network

The network is the easiest part because the answer is LinkedIn. Wasn't that easy? Don't you wish all executive decisions were that easy? It was easy because we limited the scope to CEO communications to their Professional Networks and the Social Media Network for that is LinkedIn.

Rule 2: It is about the Network Dummy!

Social Media has a basic design concept that you must work within. As Yoda said to Luke Skywalker, "You must become one with the force," and the power comes from your connections in social media. If you want to communicate, you need connections. **So, here is the rule: No network = No Voice.**

Developing Your Strategy

As the CEO, you are the tip of the spear, so you must decide who your second tier will be. The second tier varies greatly from company to company, but normally it will be all your customer-facing executives. For many companies, this includes Sales, Marketing, Purchasing, Customer Service, and Transportation. In our business, we have three market-facing positions CEO, Marketing Strategy, and Data Science. The CEO will direct the next connection request from each new contact based on messaging to the contact. The table below is a typical second-tier distribution.

Campaign Responses to Second Tier

Customers	Sales, Customer Service, Transportation
Vendors	Purchasing, Transportation
Prospects	Sales
Competitors	Channel Manager

Customers are the normal starting point for most business communications. Now we get to the grueling part, and many executives delegate or outsource this. Here are the steps:

➢ Write a 300-character long introduction for the customer.

- ➢ Get a list of all customers and contacts within the customer.
- ➢ Decide who will get the second-tier contact.
- ➢ Send the connection request.
- ➢ Repeat over and over and over.

If you did a good introduction, you should get a strong positive response from your list. Resist the temptation to use the tools that send invites to all your email contacts. This tactic uses a connection request that compromises the quality of the request. Finally, you do not want the same message sent to customers and vendors. These are different requests. Consider extending your connection requests with website content, especially for your second tier. If your approval rate of connection requests is less than 50%, consider testing new concepts for your introduction.

After a connection is accepted, do these:

- ➢ Thank the customer! – It is amazing how many skip this step, and it is cyber-rude.
- ➢ Send the names to your second tier and have them request a connection.
- ➢ As your list builds, create communications to open the dialog.
- ➢ Avoid the temptation to over-communicate and only post genuine content.
- ➢ Examine the other contacts in this business that might extend your voice.

Vendors

Normally, second in priority are Vendors, and the process is the same. Write an introduction specifically for vendors, find the contact and send the request. Normally, the connection is the Executive Contact for the vendor. Many vendor lists have the contact as the order processor or A/P person, but that is not the CEO's peer, so do not send requests to these. Lower-level contacts dilute your voice and communicate to the wrong person. This tactic takes some research, but the admin or A/P contact name will lead you to the all-company connections. Your goal is to find your peer.

Prospects

Okay, this is where it gets dicey. Some CEOs genuinely like to be at the tip of the spear, but the world of prospecting is where the world gets messy. If the CEO has a role in sales, as many do, then they should lead this process. Others support but do not get involved in sales, and those should delegate this to the lead sales executive. Here the steps are the same as customers, but it starts by segmenting the industries and writing introductions specific to each industry. Be clear about how you create value for the prospect.

Expect Higher Rejection

Prospecting is fraught with problems, and not the least of which is an extremely high rejection rate. If you select the targeting properly, have a great introduction, and a solid value for the prospect, you should see an acceptance rate of 20% or better. If it is much lower than that, test carefully and adjust until you have a good intro message. Prospects are the rest of the world, so this can and should go on forever.

Competitors & Partners

This network-building campaign is last, and you should think deeply about this before getting started. Connecting with competitors can be a source of valuable information, but it can also leak strategic information. In other words, be careful! In some cases, this might be better delegated to Channel Managers and kept out of the CEO network.

Decisions, Decisions, Decisions, Oh my!!!

In all campaigns, there are a few decisions that must be evaluated, and this section raises and explores these:

Automation – Friend or Foe?

There are plenty of automation tools around to help import your list into Social Media Networks, and LinkedIn has some powerful ones. However, with great power comes significant risk, and you need to understand this. LinkedIn has tools that will take your entire contact list from your email into the system. The problem is that people do this and send invites to people that should not be in your communication strategy. The golden rule here is do not to use a mixed list like your email contacts. If you have a clean customer list with email contacts, that could be a potential source, but again a warning: Many customer lists are administrative contacts, not executive contacts. The CEO needs to be connected to their peers. When you have a mixed list, you must compromise the connection request, which degrades your response rate.

DIY, Delegate, or Outsource

Let us face the facts here, doing this is a grind, and it is probably NOT the best use of a CEO's time, but it does need your involvement. **DIY (Doing It Yourself)** is probably the least appealing and the least leveraged way to get this done, but it might be the best way if the data is a mess. If your list requires your special knowledge to know the people you should and should not connect to them, DIY might be the right strategy. **Delegation** is a common way to get this done, but it has its challenges as well. Every CEO we know staffs the business using the N-1 formula. N-1 stands for Necessary minus one. When you look across your staff, what you find are people that have projects, goals, and missions, and giving this to them distracts from those. A rare situation is where a CEO has excess staff sitting around waiting for work to find them. If you do have that situation, meet with your HR, and fix that problem first. Outsourcing is a common way, but it has its challenges – sound familiar? The problem here is that the outsourced resource may not know your business, and with poor guidelines, processes, and procedures, they can introduce errors in the requests. The way you get this done is not important, but that you get it done is.

Audit Schedule

The world changes, and this means that on some schedule, you need to Audit your network to make sure that it is the best it can be. Because it takes a long time to send and respond to connection requests, our general recommendation is that you audit your network at least once a year. If your business is highly active with lots of personnel movements, you might want to make this more frequent.

Social Media Network Maintenance

Social Networks evolve daily, and the CEO must be trained on keeping their profile up to date. When you meet a new person, you need to request a connection. When something happens in your network, react to it, and watch your feed regularly.

Yes, there is networking after the CEO

Within this chapter, we reference the second tier in the business strategy. Each tier will require an audit and support to build out the enterprise network. The enterprise network should have a connection to every enterprise involved in the business. For example, your Marketing Department should connect to every media resource in your market, sales should be connected to every prospect, and purchasing should be connected to every vendor.

A Word about Security

Here are the steps we recommend if you decided to delegate or outsource your network audit.

- ➢ Change your password to a randomized password.
- ➢ Give it to the resource performing the Audit.
- ➢ After the audit, change it back to your normal, secure password.

This change is that you do not want to disclose your password because many times, there are patterns that people can learn to break your passwords in the future. The reason for changing it back should be obvious, so your account is not exposed after the work is finished.

Tell 'em what you told 'em

Social Media is a mystery to many CEOs, but this basic process is not. Building the right network is the key to success. Follow these simple processes, and with time and effort, you can dominate communications to your market. Ensure that you have a social media section that details the tactics and strategies involved in your business plan.

Directory Management

Directory Management is important for a variety of reasons. For some businesses, it is about referral traffic. Others it is the value of the backlink to their SEO. Some are important in one location or industry but not others. The challenge is that there are thousands of them, and getting all of them updated can consume mountains of time.

For example, if you are a restaurant, then Yelp is more than important. However, if you are a national service provider, Yelp is barely worth mentioning. The key to managing Directories is to understand the type of traffic that comes from them and how it fits your business.

Directories and Reviews

Sometimes directories and reviews get mixed, and Yelp is a perfect example of this. Yelp is a Directory and Review site and far from the only one like that. When directories host reviews, you must keep your eye on this for reviews and directory changes.

Picking Your Playground

Not every business should be in every directory, and picking the right ones to focus on is more than important. Look for some basic clues:

1. Are your competitors listed?

2. Are your competitors advertising on the directory?

3. Does your customer or prospect use the directory?

4. Is there a section for your type of business?

5. Does the directory focus on your geographical market?

If more than three of these questions are answered yes, you should consider being involved with the directory.

Pay to Play

Every directory has a way to make money, and many use a paid placement as a key part of that. However, if you pay for every directory on the internet, we can assure you that you will go bankrupt. Just like any paid advertising, you must consider the ROI (Return on Investment). The challenge is trying to get data to calculate this. The cost is easy – just ask. The value is in the traffic they send to your website, and simple math will give you a CPC (Cost Per Click) and this you can compare to your other paid advertising alternatives. The problem is, in most cases, you must test this data. Many directories will have a free listing option, but you must dig to find it. When a directory fits, it is a great investment, but they can also be blackholes where your money goes in, but nothing comes out.

Claim Your Listing

If your business is in a directory, make sure you claim this as quickly as possible. There are many horror stories of listings claimed by competitors and turned against the real owner. Getting your site claimed to a domain-specific email is your best protection from having this happen to you. A domain-specific email is when the website and the email domain match. For example, our website is www.smsrd.com, and my email address is bob@smsrd.com. That is a domain-specific email, and most directories will believe your claims over someone else without this relationship to their email address.

Advertising & Enhanced Listings

Almost every directory offers some form of advertising and enhanced listing. The question is, "Are the listings worth the cost?" and this goes back to the CPC calculation mentioned in the Pay to Play section above. If the CPC is less than the other investment opportunities like Google Ads, consider it and continually measure it.

Remember to Count Your Time

For some reason, SMBs often forget to value their time when doing the CPC calculation. Thinking time is not money is a huge mistake with directories because they are famous for consuming vast numbers of hours for ridiculously small amounts of traffic. **Remember, your time is NOT free.**

Review Management

Reviews are the basis of your online reputation, and you MUST take them seriously. There are lots of challenges created by this process within your business, and they are not simple. Businesses will get unfair reviews, and you must deal with this properly, or it can destroy your business.

Mad at Your Business

Angry reviews are what gets most businesses discussing online reviews. We have been on hundreds of calls where the client discusses what to do in this situation. The first words out of our mouth are always **calm down**. In many cases, the best thing you can do is take a deep breath and return to the topic when the venom has drained from your body. When someone takes a cheap shot at your business, the normal human reaction is to hit back, but that is the worst possible response. If you leave your emotions unchecked, you will write the best response you will ever regret. So, **again calm down**. If you know who the negative review is from, pick up the phone, call them, and settle the situation. Responding back and forth in anger might make you feel better, but it will not serve your business. Anger flying back and forth is called a flame war, and the business always loses so, **do not engage**.

Loves Your Business

Love your business reviews are what everyone seeks and when you get one, you need to be grateful and magnanimous. Step one in all complimentary reviews is to say thank you. What you say in response to a review is viewed by many more people than just you and the reviewer. This public exposure is the power and the pain of a review.

All the Rest

The one-star and five-star reviews are the extremes, but most reviews are between these two points. They are the reviews that many people value more than either of the extremes. The mid-range reviews tend to be the most honest and the most valued by others. The key to these is to resolve any issues raised and to thank them for the review. This follow-up will often result in a higher rating because you paid attention to their pain and showed that you cared.

The Difficulties of a Review

The biggest problem with reviews is getting them. If you think you can just sit back and wait for the reviews to flood in, you are painfully wrong. Getting reviews takes a deliberate process of asking for and following up on reviews. If you do not ask, you will not get many.

Next is knowing that they exist because there are thousands of places that a review might live. The big sites for reviews include:

1. Google My Business
2. Amazon
3. Facebook
4. Yelp
5. Trip Advisor
6. Better Business Bureau
7. And many others

If your business manages lots of reviews, there are some great software solutions to monitor and respond quickly.

Facebook

Facebook is the 800-pound gorilla of consumer-based social media networks. With an estimated 2.4 billion users each month, the numbers here can be staggering, but that is not important. What is important to a local service SMB is the local users that use your service. Many of these tell their friends about every little detail of life, and comments spread like wildfire.

Check Your Ego at the Login

Rule 1.01 in Social Media is never, ever, never respond when you are upset, angry, or **otherwise emotionally compromised**. It is easy to get wrapped up in a complaint and respond with the **best reply you will ever regret.** Consumers often go to social media and vent their anger in unfair ways, and you must just take it. To respond in kind can start a flame war, and the businesses never win. This sort of angry tit for tat exchange is not just between you and the other person. The universe is listening! So, calm down and offer to resolve the problem and take the conversation offline as quickly as you can. Do not use social media or email when dealing with an upset customer.

Remember this is PUBLIC

I do not know why this concept is lost on so many, but it is. As a business guideline, we would like to give you this:

1. Never post something online that you would not want a lawyer to read back to you in a courtroom.
2. Never post online something you would not want your next prospect to read when deciding if they should call you.
3. Never post something online that you would not put on the homepage of your website.

You want to project a professional image that supports the image of your business you want in the market.

Facebook has an extraordinarily strong advertising system that is worth evaluating for branding purposes. This traffic quality is more like the Google Display Network, not the Google Search Network. Generally considered more of branding rather than direct sales strategy.

Yelp

Yelp is a strange place, and many businesses have had problems with this network. **Yeah, it is not just you.** Yelp often attracts  people when they want to vent a complaint. So bad reviews are known to happen here, and like all social media networks, the business has no rights or fair process for rebuttal.

In a local service, SMB Yelp is something you must watch carefully. When a complaint comes, respond in public, and move the conversation offline as quickly as possible. Say things like:

"We are sorry we did not reach our goal of providing you with an exceptional experience. Please call me at 800-555-5555, so we can make this right."

Then do what you have always done and find a way to make it right. You might never get the person to change their negative review, but you can make it better.

Like all social media and search engines, the algorithm that decides who goes first in these listings is not easy to understand. Yelp will throw out some reviews, and they always seem to be the good ones, but nobody is going to change that for you. We have found that if you are an active advertiser that Yelp seems to respond to that. They shouldn't, but it appears from our observations that advertisers get preferred handling.

Yelp is an aggressive marketer of their advertising, and we have noticed a correlation between bad reviews and Yelp sales calls. It seems like one might trigger the other ☺. We have seen situations where bad reviews go away when the advertiser starts to spend money with Yelp. We are not saying this is a cause-and-effect relationship, just merely an **observation of correlation.**

LinkedIn

It is not what you know; it is who you know. While our world is constantly changing, this adage rings as true as ever in Social Media. The volume of your social media voice is created by the quality and quantity

of your connections. Without connections within the network, you simply have no voice. Let us explore the details of how to improve LinkedIn presence for business-to-business customer-facing staff.

Business to Business Customer Facing

This tactic is specific and assumes that you are building a professional network because you hold a customer-facing position within the business. Examples of customer-facing roles are sales representatives and executives with active roles in sales. Since LinkedIn is the dominant professional network, we also assume that the business network you seek to build is B2B. LinkedIn certainly has consumers in it, but at its core, it is a professional network. To run into this network with a consumer tactic is a marketing disaster looking for a place to happen.

The Connection Request

If there is one thing that separates winners from losers in LinkedIn, it is their connection request. A connection request is sent to the person you are saying hello to and is limited to 300 characters and spaces. A good connection request has these components:

➤ It is Personal
➤ Explains who you are
➤ Explains value to them
➤ Strong "Call to Action"

If you hit all four of these, you will have a **high "Yes" rate**, and you will start a relationship with a positive first impression. If you think that writing this is easy, **you do not yet understand the challenge**. On average, you have 42 words to project your personality, explain who you are and why a relationship with you is mutually positive. Doing this in 42 words or less is a huge challenge. You will know that you have done a respectable job with this when your **"Yes" rate exceeds 20%** from a cold contact. Since we recommend driving this by the title of the potential contact, getting personal is possible, and it sets the stage for who you are and how they get value.

Title Search

The foundational tool for building your network is the search and, more specifically, the title search. Our goal is to connect to a specific level within targeted organizations. Here is an example of a title search:

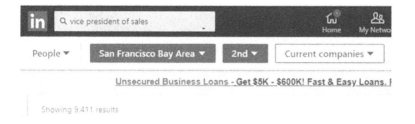

Showing 9,411 results

In this example, we are looking to connect to Sales Executives and Vice Presidents of Sales. These are two quite common job titles that are typically utilized with this tactic. There are some challenges to watch out for on this. While this shows a result of 9,411 results, you cannot get to all of them. At most, LinkedIn will give you 100 pages with ten people per page for a total of 1,000. If your search exceeds 1,000, you cannot see the other 8,411 listings. These are only my 2nd level contacts. If we add 3rd level contacts, the number jumps to 68,171, which increases this problem.

So how do we get around this limit?

The same way you eat an elephant: One "byte" at a time. The solution takes time. You must work the list little by little until you get through the first thousand results. As people accept your connection invitation, they move to your 1st level creating an opening in the top 1,000 2nd level. This process can get complicated very quickly, so ensure that you keep good records of your requests. LinkedIn will limit your daily requests to 100, so processing the top 1,000 will take ten days. The key here is a continual effort at a reasonable level.

Other Limits

LinkedIn also has a cap on the total pending connection requests, which was recently increased from 3,000 to 5,000. It is a must to remove stale requests to create new space in your account. Click on your network tab, then Manage all on the right side:

Then click the **Sent** tab

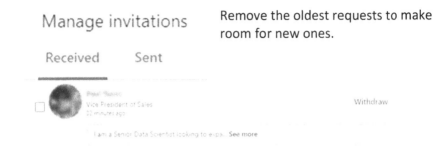

Remove the oldest requests to make room for new ones.

What happens as you work the account is that a certain percentage of people will accept your connection request. The connection approval moves them to a 1st level contact, changing the sequence of all the lists. If you continually work on your network, it will grow, but don't go too fast because you will be flagged as a connection spammer.

It is about Quality, not Quantity

Businesses have long known that it is not what you know but who you know and how you know them. In social media, in general, and LinkedIn specifically, creating a connection is equivalent to saying hello in the physical world. If the person says hello back, then it is an invitation to have a conversation. From conversations come relationships, and from relationships, you get value.

LinkedIn Tries to Help

If you click on the My Network tab on your account, you will get this:

These are the recommended connections from LinkedIn based on your account, and honestly, they do a respectable job. With just one click, the request is sent. How cool is that? Be careful, though, because with great power comes significant risk. This tool sends a standard "Please join my network" message, a clear sign that you gave no effort toward the relationship. These requests will get rejected at a much higher rate than those sent with a personalized message. We use this function from time to time, but if you click through to the profile and use a custom introduction. It shows you care, and it will be accepted at a much higher rate.

Business Versus Personal

Ownership is a hotly debated issue globally, and we do not have the answer for your situation. Our goal is to help you understand the issues. Businesses invest in building networks to communicate with their market, but LinkedIn profiles are based on a person. So, who owns this business asset? If you are creating a new profile and get a written agreement in advance, there is a possibility that your business can own it. If you hire a person with an existing profile, the business claim to ownership is beyond thin. We have not seen this evaluated yet, but it is a conflict that is hard not to see. We are **NOT A LEGAL resource** but, it stands to reason that if a business invests thousands into building a profile, they would like to own it. Our best marketing and business advice are to understand the risk you are taking clearly. Talk to a good business lawyer, not your marketing agency, and get an agreement in advance, not at the end of the employment relationship.

Do, Delegate, or Outsource

Work gets done in one of these three ways, and each has its strengths and weaknesses. As it applies to building your connections in LinkedIn, you can do it yourself. This chapter gives you the basics of how and certainly nothing here is rocket science. If you read this far, we are sure you can **do the work**. The question is, should you? Most customer-facing staffs are professionals and executives, and honestly, the process here can be a bit mind-numbing for people of that skill level. You can **delegate,** but that can spread your staff too thin. Another issue is that social media changes a lot, so keeping up to date is a challenge for internal staff. The exception to that is a delegation to an internal social media management department. **Outsourcing** often makes sense, but it is not without its challenges. In an outsourcing relationship, make sure they never try to represent the business. In an unusual response, they need to immediately raise it to the profile owner and get out of the loop. Most outsourcing resources will perform this task full-time allowing them to keep training up to date while staying within the limit imposed by LinkedIn. Done properly, this is one to two hours a week.

Goals

Utilizing any tactic requires goals and a budget. We recommend that you start at a pace of one hour per week. In one hour, you should be able to reach out to 100 people and get yes responses. You should get a 20% yes rate, but something is wrong if you do not get these results. Examine your message and your target and make the necessary adjustments. After 30 days of getting your 20%+ yes response, you can ramp this up to five hours. At that point, you will start running into the system limits for network building.

Budget

To build a network, you need more than the free account on LinkedIn because your search results will be limited after a few searches. An annual Premium Business account cost is $575.88, which translates to $47.99 per month. The labor cost depends on how you value your time. For this discussion, let us say your opportunity cost is $100 per hour. This labor adds $5 to your per connection cost. Compare this to most outreach efforts that create permission to communicate, and you will see that **LinkedIn is a bargain**. Opportunity cost is the value you could create in your business if you were focused on something else.

Do, Delegate, or Outsource

LinkedIn Network building is often delegated or outsourced because customer-facing individuals have extremely high opportunity costs. Delegation sounds good and seems to cost less, but it is often difficult to manage and maintain. Few businesses have someone sitting around waiting for work to do. This work often gets dropped for higher priority work, and it takes time to keep up to date with changes in the network. Outsourcing is a solution that typically brings in lead costs in the $5 range.

Now that the simple stuff is done

While in the beginning, building your connection network might seem like a daunting task, it is just the first step of a long journey. As you build out your network, you must start communicating with everyone and create your professional persona.

Email

Email is one of the most powerful tools in your Marketing Toolbox. It takes time and treasure to build good lists, but the rewards can be amazing. Email allows you to reach your customer base at an exceptionally low cost. Some people will say that sending email is free, but that is NOT TRUE. When you send an email, you incur the cost of losing people from your list and the cost of replacing them.

Email Rule 1

Never send an email that does not contain a value for the person receiving it. Consider first the value proposal of each email you send and the person you are sending it to. It is amazingly easy to become an email pest, and the reaction you will get is not good.

Email Rule 2

Have a plan and execute it consistently. Remember that you are painting an image in your customer's mind, and every touch is one more paint stroke on your masterpiece.

Email Rule 3

Track everything you can. Make sure that all links on the email are tracked and break your content up to measure response. Do not put entire articles in email. Put a teaser copy and a tagged link in the email.

Building Your Email List

Building your email list is a grand challenge that is won with a steady grinding toward the goal. There are a few simple things you can start with.

Ask every person you serve:

Would you like me to send your invoice copy via email or text? Then make sure that email makes it to your list.

Email Cost More Than You Think

Many people mistakenly think that email is free, but that is wrong. It costs money to build your list, and every time you send an email, the person at the other end might unsubscribe or, worse, register a SPAM complaint with your ISP. For fully qualified opt-in emails, we have seen costs up to $10 per email. When you send a batch of emails, take the number of unsubscribes times the cost to procure, this is your postage cost. We know many companies that have tens of thousands of dollars invested in the building of their lists.

Use it or Lose it

Email lists are not a fine wine, and they do not get better with age. Lists decay and too much decay will make it difficult or impossible to restart. Studies have shown that about 2% of your list will likely decay each month. This math shows that your wonderful performing list will be rot from the core out in 4 years. If your bounce rate or spam rate is too high, you run the risk of being classified as a spammer, and that can follow your business around for years. In the worst cases, you can get blacklisted and not be able to send anything to anyone.

Conditioning the List

Sending a list on a regular schedule and keeping up with the maintenance will develop a powerful list. By removing the 2% that dies each month, you will avoid getting into the range that causes blacklist problems.

Not Read or Dead?

Knowing the condition of the email list is a BIG challenge, and the reported read rate is wrong. The Read Rate is the number of emails opened divided by the emails sent. The problem is that the tracking of email opens is not perfect. When a service like Constant Contact or Email Chimp reports that 15% of your emails were opened, this means at least 15% were opened. However, we have seen lots of well-documented cases that show that this data is 50% or more in error. There are lots of ways to break the tracking, and there are few opportunities to overreport. The errors come from many sources, including firewalls, privacy software, and other things trying to perform some other function.

Invest to Refresh

Email is a powerful marketing tool, but you must continually build your list. As a minimum, you must replace the 2% you will lose each month. It is that simple. If you have a list of 1,000 emails, you need 20 new emails just to break even. If you want your business to grow, it needs to be more than that.

Pick the Low Hanging Fruit

Every time you send a batch of emails, you will get a wealth of information back. Some of that data can reduce the number of emails you have to replace. Many emails are simply being updated, and you will get an auto-reply email back that will tell you the change you need to make.

Dialing for Dollars

The phone can often be the best way to build an opt-in email list. Ideally, we would like to get subscriptions from the website, but that has not worked for most businesses for years. Often the way to build this is to call your prospects and ask their permission to send them an email. If you have an excellent value proposition, many of these will say yes. With a qualified list, we have found that a 15% yes rate is possible with a call rate of 15 calls per hour. Your number will vary based on the value proposal in exchange for the email permission.

Role-Based Emails

Role-based emails can be good or bad, depending on your specific situation. Examples of role-based emails are something like info@, sales@, admin@, and more. These emails are not a person but someone assigned to manage a specific flow of incoming emails. If your email has limited value for the reader, we can assure you that this person will delete your email. When you port a new list into service, most of these emails will be removed from your list before importing.

Phone Training

Next is to train the person that answers your phone. Getting a text number or email address should be part of every exchange you have with the market. If you get ten inbound calls and nine emails/texts, then find out why. Phone training is complicated, and you must balance the goals of the call but getting this data is critically important.

Go Back Through Your History

If you have been in business for any amount of time, then your prior job tickets can be a treasure trove of marketing data – so Mine It! Start building a list to serve your business in one of three ways: email, text, or snail mail. Yes, we said snail mail. It might be more expensive but reaching your market at a cost-effective level is the goal.

Attend Local Events

Every community has events that attract large groups of people, and these are a major source of outreach to a community. Attending things like a Chamber of Commerce mixer can be great for business. While this is a business-to-business event, we can assure you that many people are consumers and might need your products or services in the future. If they have met you personally, they are much more likely to call you than a stranger. So, get out there and shake hands. Gather business cards and start communicating with them.

Drive Blog Traffic

One great use for your email is to drive recurring traffic to your blog. A cycle of an article, email blast, and display advertising is a fantastic way to build a strong following.

To Blog or Not to Blog

Every Internet Marketing Expert that is worth talking to will tell you that **"Content is King."** What they fail to tell you is that it is also challenging work. We cannot tell you how many meetings we have been where someone, normally the highest-ranking person in the room, will state, "We want to rank first for this list of keywords." After which, they produce a list of 50-100 words. Then tell them that they are not important enough to earn 50 keywords and they lack the content earn more than one or two. We try to do this with respect and finely polished language, but ultimately the statement must be made.

I want to rank first

Think about the absurdness of this statement within the context of most websites. They want to rank first for words that do not even exist on their site. They have only a few pages of content and what content does exist is 100% sales pitch.

If they genuinely want to compete for the keyword list, they must have at least one page for each keyword, not four, not three, not two, but one to one. The page needs to answer the question of the person that searched for the intent of the word. To put this in Google jargon, it must be high-quality, relevant content that answers the search query. If Google judges it to be the most relevant answer to the question, it will win the first position.

Now play the Blog card

Your website blog comes into the strategy to address the page content discussed above. Start with your wish list of words in priority sequence and then write an article about that word. Then place that article where it can be found and move on to the next word. Do this at a pace somewhere between 2-4 times a month for a few years, and you will be on your way to earning that traffic you wanted.

What is wrong with this?

Blog articles, like most content, age poorly. If you start with a list of 50 keywords and a two article a month strategy, you will finish in about two years, just in time to start over. Each article must be optimized for one keyword and user engagement aligned with your marketing messaging. These articles have to be high-quality because it is of no benefit to your business to get people to the website and embarrass your business with poor content.

Blogs need a little help from their friends

Blogs often need help from advertising to get started. When your content first hits Google, it will get some points for freshness, but that can quickly fade. As you are building the reputation of your blog, it helps to push primer traffic there. Primer traffic is early traffic to engage your audience and is commonly PPC traffic from search or display. Blog content that is done properly will engage your paid visitor, and they will explore your site to learn more about you. We call this an engaged reader, and it is an early indicator of value being created for your business.

Guest Blogging

Once you have perfected your blogging skills, consider reaching out to industry-related websites and asking to be a guest blogger. If the site is open to this, they will probably require that the content you publish remain unique to their site. If the site reaches out to you, be careful because credible sites do not have to do that. Most that reach out will either want to charge you for publishing or require some advertising purchase. Either way, watch your wallet.

The Secret to Blogging

The worst kept secret on the Internet is that a successful blog with a consistent flow of high-quality content is the secret to owning the top SERP position. A blog takes time and effort over an extended period, but it is the way to build consistent business over the long term.

Distribution Design

You build your lists and drive messages out to create leads for your sales pipeline. Your Distribution Design is what facilitates that process. This section applies to traffic other than emergency search. The goal of your content distribution is to build a multi-step marketing process that will bring business to you in the long term. Normally the system will consist of:

1. Blog
2. Email List
3. Snail Mail List
4. Social Media
5. Advertising

Using the right tool for the job

The blog is the content depository and the base of your distribution system. The blog is where you want your traffic to go, and the content is your on-page website optimization. The informational needs of your customers should drive the articles built here, so listen carefully.

The Email List drives traffic to your Blog. The idea here is to send a reminder email to prior customers and prospects with teaser copy about your latest content on the blog. For the system to work, this needs to happen regularly and frequently enough that your brand is refreshed.

For an SMB, we recommend that you publish at least one article per month. Enough to refresh your brand in their mind but not so often that you become a pest.

Snail Mail, also known as regular mail, is expensive but powerful. You can use this to drive blog traffic, but it is best to have a specific offer and an article promotion because of the expense it is best. Snail mail rarely passes the common-sense cost test for traffic only.

Snail Mail is a solution to the neighborhood targeting strategy, and here is how that would work. Every time you have a job in your target area. Look at the other homes in the area and if they look like your target, then put them on the list. All you will likely get is the physical mailing address, but it is a place to start. From there, you can say – we just helped your neighbor Bob with his water heater, and we would like to be your plumber as well. Maybe we can look at your water heater, so you do not end up with an emergency plumbing call.

Social Media is another tool that can push traffic towards the blog. Again, this is done with teaser copy, and most of the social media is done with visual assets. An interesting picture that draws attention with a solid call to action is the key to this.

Advertising is the last on the list, but it can be one more powerful in many cases. Advertising is run with display network placements such as GDN, LinkedIn, Facebook, and others. This content display means that the person runs into the offer without initiating a specific action seeking out the content. They did not search for it; they just ran into it while looking at other content.

Now let us put some details around this, so it all makes sense. Let us say that you are in an area where water quality has become a hot topic. The city publishes a water quality report, as they are required to do. In that, the City discloses certain facts. Here is what that might look like.

As a smart marketer, we do the same thing that a local reporter would do. We find the story and then tell it while connecting it to your solution. This water report discloses problems with Nitrate and Hardness. As a professional plumber, you know that a RO unit can help with this. So, you create a distribution cycle with:

> ➢ Blog Article Discussing the problem and solution
> ➢ Email Teaser promoting the article
> ➢ Snail Mail Flyer mailed to your hyper-local target market
> ➢ Social Media Posting
> ➢ Display Network Ad to promote the Article.

As a marketing expert, we would modify this to show family and friends on the pure water side of the diagram. We would make additional points on the improved taste and safety of food prepared with filtered water. As a business, we realize that installing a RO unit gives me a perfect path back to this customer every year.

You can do the same thing with almost any product or service. New products like wireless moisture monitors, water use monitors, and many others provide an unlimited number of these opportunities. Once you have built out one idea for each month in your marketing plan, you can then redeliver that cycle for several years.

Content

Content for plumbers is always a tough conversation because of the nature of the plumbing business. Many think that plumbing is a boring topic, and they are right but not all the time. Plumbing is interesting to people when it is important to them. So, here are some starting ideas for your content plan.

Listen to Your Customers

Your customers are the priority when it comes to content generation. What do they ask you? What are their issues? Where is their confusion? How do you answer these questions now? All of these are important questions, and just actively listening will get you there.

Advice on how to avoid plumbing problems shows you care, and the fact that you care is what separates you from all the others in your market.

Monitor Important Community Issues

Almost every community in the US has a water report, and one community idea would be to review and comment on that. If readings are getting high in the community, you might want to offer some water treatment ideas. An RO unit or home water treatment system can be a great idea for those concerned with their water.

Life Spans

The life span of products creates a demand event, and that is when you want your name at the top of their thinking. Especially if you offer tips on extending the life of expensive items, an example of this might be a reminder sent to your lists that says "Time for the annual flushing of your Water Heater" read how here. That takes it to a blog page with instructions on how to flush a water heater. Just by accident, it has your annual plumbing maintenance service that includes a full inspection and a flush.

Advanced Special Topics

This section contains chapters that do not fit cleanly into a primary section of the book but are important for understanding.

Machine Learning or Human Experts

Advertisers are at a crossroads on this strategic question of automation vs. human experts. If you think you know the answer, then you do not yet understand the issue. Google's position is that Machine Learning is the better answer, but their agenda is to serve Google's best interests, not yours. Our goal today is to start the discussion and push the debate forward.

Artificial Intelligence, or AI, is an umbrella term with Machine Learning, or ML. Nobody can deny ML has been making progress over the last several years. The inventory of algorithms grows longer every day. There are plenty of people throwing the term AI around, and the reality is nobody is even close to reaching true AI in advertising. There are many definitions of AI, and many of them cross over to Expert Systems and Machine Learning. True AI would have to be capable of innovation, not just incremental improvement, a level we have not seen yet.

Machine Learning

Machine Learning uses data and algorithms to learn. It can produce impressive results in some situations and find patterns and relationships people would miss.

Machine Learning only has historical data patterns to react to and cannot predict the future. For example, a pandemic has a ripple effect that changes life overnight and is a major driving factor towards changing market trends. A human set of eyes can adjust for this and make necessary changes much quicker. A machine will fall behind in seeing that things changed but will not understand why.

Due to this and many other reasons, this is not a utopian solution, especially with Google slanting the playground in favor of Machine Learning (otherwise known as cheating). As Google moves through implementing ML, they are taking data and control away from the Human Expert. In simple terms, they are cheating to make sure that ML wins.

As Google rolls out ML, they are presenting studies and conclusions that are highly suspect. Testing outcomes are difficult because it is impossible to recreate the same situation. They evaluate across a wide section of data, which is a good thing, but it does not necessarily apply to your situation. ML is seeking patterns in the past and projecting them into the future, which does not always happen. A classic example is an ML routine with seasonal data, so it over projects into the off-season. If your business always repeats its performance, ML is your tool, but if unexpected things happen, you must be careful about following ML.

Human Experts

Human Experts are not a utopian solution either. People cannot digest data at the pace or volume built today, and they often miss data patterns and relationships. Their memory is imperfect, and their actions are inconsistent. They do have a much broader view of the world, and Human Experts are capable of intuitive logic leaps creating innovative opportunities. These two factors combined give Human Experts a major advantage over even the strongest machine learning tools.

Machine Learning versus Human Experts

If you are looking for a quick winner or loser, this is not the battle you want to follow. Both strategies have their advantages and disadvantages, and the winner will not be clear for an exceptionally long time, if ever. The key to this battle will be when you play each of these in your business strategy. If you segment your challenges into small cells with tight definitions, then picking the right tool will become more visible. The bigger and broader the marketing cell, the more the Human Expert skills become valuable.

Google's system-wide ML implementations have a fatal flaw. When implementations are available to everyone, then you do not get an advantage. You get average. You get average because everyone is using the same tool. The study that showed the 5% improvement in the measured metric is gone when everyone has the same tool. The initial advantage will go to the early adopter, as it does now because they will get the short-term gain before everyone else jumps on board. We have long used an entered early and measure carefully tactic. We follow this with a quick exit when the performance starts to decay.

Incremental Improvement or Innovation

One area that is clear in the fog of sales is an improvement versus innovation. If your sales goal is a 20% improvement, then the power of ML is a good bet. If your goal is more than 50%, put your money on human experts. ML is not designed to make intuitive leaps, but it is great at squeezing out incremental improvement.

Our Prediction

We love predicting the future and then coming back ten years later to see how we did. This paragraph is purely my opinion based on my life experiences. We predict that the winner of this battle will be a hybrid of Machine Learning and Human Experts. ML will evolve into high-performance strategic tools, and human experts will become strategic tool users. As human experts, we will configure and tune the ML to exactly fit the business. With the human experts relieved of the painful data analysis, innovation will speed up even more than today. Google will give up on finding the one grand solution and become an ML Tool Vendor providing a trusted privacy shield to the end-users.

Data Reveal or Conceal?

Data is the center of our universe, and we are always seeking actionable business intelligence from it. Data does not always reveal its mysteries and secrets easily, often concealing its intel using sleight of hand. Nobody in business designs data to conceal business intelligence, but it happens. Let's look at a story that demonstrates unintended data concealment.

Several years ago, a high-end luxury brand came to us to be part of the team for a digital campaign launch. Our role was as the programmatic placement and data experts, specifically for the Google Display Network. The creative team designed great content for placements, and we placed them in the Google Display Network using a variety of tactics, including keywords, named placements, audiences, remarketing, and others. After checking everything multiple times, the launch date and time were set.

On launch day, we activated everything, and within a few hours, the impression levels were building, and we started to watch the data closely. Over the next few hours, the team went from concerned to happy. Within a few days, impression levels were in the millions, and the click-through rates and orders were solid. The client was pleased, and there were lots of high-fives and nice-to-read emails. There was some basking in the glory of a job well done. My gut instinct was on edge because we were well above plan, but everything was coming up roses, so we went with it. I remember thinking that it seemed too good to be true, but sometimes you get lucky in marketing.

You knew there was going to be a 'but,' and here it is. Over the next couple of weeks, the results faded, and smiles turned to frowns. The impressions and clicks held, but the results went from great to barely making the numbers. This turn in the data sent our team scurrying back to see what was happening. As we peeled back the traffic, we quickly found data that concealed the reality from our eyes in the early days. Under the display traffic, a layer of referral traffic turned out to be the source of many orders. Referral traffic is almost always low volume and excellent quality.

As we researched this, we found that six months before the launch date, a PR firm had pitched a story to a large industry publication. This PR firm no longer worked for the client, but the article was already in the pipeline, and the publication went to press with it on the launch date. The great results were purely an excellent accident. The traffic it generated was low volume but super high quality. The traffic barely moved the needle in the inbound traffic, but it was rich with orders. What everyone saw was a jump in inbound traffic from the campaign with a spike in orders.

The lesson learned from this is never to stop drilling into the data, just because what you see makes you happy. Human nature is to take the win, bow, and take the glory. We looked at the start and finish, and we stopped because it was the answer we wanted. The golden rule here is to continue to follow the data until you get to the end of the results.

Marketing data is notoriously complex, and everyone wants a clean, simple path from start to finish. Unfortunately, that rarely happens with sales paths that involve 5 to 20 touches along the path to the close. Therefore, we must watch data at both the micro and macro levels within the sales paths.

Being Hyper-Local

There is no doubt that being hyper-local is a good strategy for many businesses. When we first start working with a location-driven business. We always have the conversation of their targeted geography, and honestly, it is often a weird conversation. In this targeting discussion, many businesses have the desired reach that greatly exceeds their ability to execute. Let me give you an example. I grew up in Anaheim, so I know the geography well. Recently I spoke to a plumber with three trucks who defined their geographic target as North on the 57 to Pomona, South on the 5 to Irvine, West to Cypress, and East to Anaheim Hills. Here is what that looks like on a map. This target might seem reasonable to some, but it greatly exceeds their ability to serve and not by a little bit. To estimate a market reach that makes sense, you need a few numbers.

1. How many households in your market? 100,280
2. How many calls per household in a year? 0.50
3. How many calls per truck per month? 84

If you multiply households by the number of service calls per household, you get the market potential. Take that and divide that by the service tickets needed to keep the trucks busy, and you have an idea of what you are dealing with. Plumbers are the example here, but this same logic process can be applied to almost any locally targeted business.

Back to my Anaheim example with data from the US Census Bureau https://www.census.gov/quickfacts/anaheimcitycalifornia

Anaheim has 100,280 households. The estimate of plumbing service calls per household is one per two years. The math works out to 4% each month (100% divided by 24 months = 4.1%). For service calls per truck, take the daily goal of four tickets times 21 workdays, and you get 84.

This example is rough math, and there is no need to get exact because it is an estimate. Anaheim will generate 4,178 service calls a month (100,208 households time 4%), and with one truck rolling, you need 84. Math tells us that in a perfect world, a condition that does not exist, Anaheim has enough business to keep 50 trucks busy. This guy had 3.

Suppose this business was in Anaheim proper (the east side of the map). We would recommend they focus on that market first and dominate that. The potential of this market is 50 trucks, so assuming they have less than ten, this market is plenty big enough.

Being hyper-local reduces the transit time and increases the profit of the Plumber. In a case like this, focusing resources on developing the hyper-local business could improve the number of service calls per day and move the needle in a big way.

Most Plumbers target way too big of a region and should think about the long-term value of that. We do not propose that you turn away business outside of your target, only that you should target your marketing investments to be hyper-local.

Math is fun...let us do More!

The first number we used for estimating the local market was from the Census Bureau, but there is more data available in most markets. Here is the housing breakdown for Anaheim. Many plumbers focus on Owner-Occupied homes, and for a good reason.

Table 2-15

Estimated Unit Size by Tenure, 2011

	Owner-Occupied		Renter-Occupied		Total Occupied Housing Unites	
	Units	%[1]	Units	%[1]	Units	%[1]
Studio/No Bedroom	26	0.1%	1,650	3.1%	1,676	1.7%
1 bedroom	792	1.7%	15,549	29.4%	16,341	16.6%
2 bedrooms	5,382	11.8%	25,561	48.3%	30,943	31.4%
3 bedrooms	21,148	46.4%	7,422	14.0%	28,570	29.0%
4 bedrooms	14,823	32.5%	2,473	4.7%	17,296	17.5%
5 or more bedrooms	3,445	7.6%	315	0.6%	3,760	3.8%
Total	*45,616*	*100%*	*52,970*	*100%*	*98,586*	*100%*

Notes:
[1] Percentages may not equal 100% due to rounding.
Source: 2011 American Community Survey B25042.

We can see here that 35,971 homes are single-family 3- & 4-bedroom homes that are owner-occupied. For many residential plumbers, this is the prime market, and it is estimated that it would take 17 trucks to service.

Hyper-Local Targeting

When running Google Ads, location is one of the bid modifiers. You can, and should, use that to your advantage. Make your office the center of a radius and then go out 3 miles with a higher bid, then 5 miles, then the rest of the market. Since the commute home will be shorter, your profits will be higher. By modifying your bid for the traffic closer to your office, you can dominate those searches. For example, the inner circle would have a modifier that adds 20% to the bid, with the outer circle getting 10%. If your business provides service driven by the location, then being hyper-local must be on your agenda.

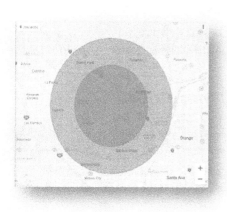

Train for Phone Calls

SMBs spend an inordinate amount of time and treasure to get the phones to ring. We have audited calls many times and cringed as we listened to recorded inbound calls and how they were managed from time to time. A simple rule here is do not hesitate to train those who answer your phone.

They can hear you smile

Projecting a friendly personality on the phone can be the difference between a new customer and a lost opportunity. In many cases, businesses pay hundreds of dollars to get that inbound call to waste that money by having the wrong person answer the phone. Many professionals in this field keep a mirror on their desk to remind themselves that people can hear you smile. A smile is the strongest influence in a personal exchange, and saying that it can be the difference between success and failure is not an overstatement. When you smile, it changes the tone in your voice, and people react to that.

Drinking for Dollars

Drinking water can improve sales performance. Don't believe me? Try it! What do you have to lose? Drinking water changes the way your voice sounds, and it's good for your health. If you want to explore the details, read this article:

https://www.ncbi.nlm.nih.gov/pmc/articles/PMC2925668/

Of all the ideas for improving sales performance, this one is stupidly cheap. People need water, so keeping it around is good for business. Most doctors highly recommend drinking water for all sorts of reasons.

Stand Up for Sales

When you answer the phone – STAND UP. Standing up changes the way your voice is projected, and it improves sales! The cost for implementing this is zero point zero dollars. So as business improvement ideas go, it has a great ROI.

Have a Plan

The person answering the phone must quickly classify the type of call and plan how to move that call to the proper conclusion. Emergency calls must convince the caller that you will solve their emergency problem quickly. We know that people will stop shopping only when you reach this standard. If the person is left with any doubt, they will move on to your competitor. Write out your phone plan and brief every person that might answer the phone on how to move through the plan.

Monitor, Listen, Coach

If you follow the recommendations in this book, your calls are already tracked and recorded for training purposes. We highly recommend that you randomly Listen to and Coach staff on how to manage the calls. If you do not have this expertise, then reach out and hire a phone coach. A quick search on the internet will give you dozens of options to explore and find the one that fits your business. Here is an example:

https://www.callsource.com/call-coaching/

https://www.callsource.com/4-stages-of-call-coaching-to-boost-appointments-infographic/

This article proves the value of this training. I know the link is crazy long, but you can also go to callsource.com and search for plumbing.

https://www.callsource.com/blog/how-one-hvac-and-plumbing-company-added-over-40000-to-their-bottom-line-in-3-months-with-call-coaching/ The bottom line is that you pay big money to get the phone to ring, but you only get value if you answer it properly.

Call Centers

Call Centers can be a great solution for many SMBs, but you must select them with great care. Interview them as you would a prospective employee and make sure they can meet the standards you want for your business. They must be scripted, and you cannot expect them to become an expert in your business. Within the script, you must provide points to bail on the call and get the call to an internal expert.

Ask for Referrals

Referrals are sales & marketing gold, and everyone loves them. They cost less to generate, and they close at astronomical rates compared to all other lead sources. The challenge with referrals is that they do not scale. You cannot just turn up the budget and get more referrals. The challenge is that referrals happen when a friend of the referrer needs what the referrer thinks can be helped by your product or service.

Everyone loves referrals, so what are you doing about it? Here is a simple idea. After each service call, send the customer a thank you note along with three business card-sized serial numbered coupons. If a new customer claims the discount, give the initial customer a check and three more cards. Because you just found yourself a genuine real-life evangelist for your business. Repeat this process with every new customer, and over time this tactic could be a significant percentage of your business. On top of that, it just might be the most profitable marketing investment.

What Works

In every successful and scalable referral campaign, there are two common attributes. (1) Ask early and often. (2) Give Referrals. Only the strength of reciprocal influence will move customers to a referral.

Voice Search

As a kid, I thought how awesome it would be to talk to a machine as Captain Picard did with the Enterprise. Google, Apple, and Amazon are racing to make this a reality. So far, Amazon is closest with the introduction of the Echo-Auto, but our focus here is not to discuss products. It's how to be found during a voice search.

What is Voice Search Used For?

When it comes to business, voice search is primarily used to find a local service, business, or product to buy online. For local searches, getting directions is typically the next step. If you are in Home Services (i.e., plumbing), proximity to the user could be a deciding factor.

How does this apply to your business?

The market shows that people are quickly adopting voice search, with research showing that 41% of adults use voice search at least once a day. Trends show that at least 50% of searches are done by voice in 2020. With half of all searches conducted via voice, your business better be ready to be found!

How do you make sure your business can be found from a local voice search? To answer this, you must understand where each assistant gets its information and what share of the searches each one has. Check out the table below:

ASSISTANT	SOURCE	MARKET SHARE
Siri	Apple Maps Connect supplemented with Yelp	36%
Google	Google My Business	36%
Alexa	Yelp & Yext	25%
Cortana	Bing Places for Business & Yelp	19%

Notice that Siri and Google Assistant have the majority of the Usage Market Share (36% each), followed by Alexa (25%) and then Cortana (19%). When it comes to in-home devices, Amazon has had much of the market share, but the number of deployed assistants includes smartphones.

Claim/Create Your Listing

The first thing you need to do is claim or create your listing in each source. The first ones you should tackle are Google My Business (GMB) and Apple Maps Connect since these have the largest usage, followed by Yelp and then Bing Places for Business. Now that you have accurate listings, you're all done. I was just kidding! This step only makes you eligible.

Optimize Your Listings

You must optimize your listing. In other words, you must accurately provide all the information that these listings ask for and keep it current, including the absolute correct version of your URL, including the HTTP, HTTPS, and with or without www.). On the same note, only provide useful information. For example, only provide an address if customers physically visit your business. For businesses that serve customers at their homes, there is an option to indicate your service area instead.

Other common optimization considerations include photos and reviews. Upload good quality photos of your business, including employees doing work, your logo, interior & exterior photos of your business. **Do Not Use Stock Photos!** The photos uploaded should accurately represent the business because people can easily tell which photos are authentic. Furthermore, stock photos make your business look corporate and not local, sending the wrong message to your audience.

When it comes to reviews, it is essential to respond promptly. It is just as important to respond to legitimate bad reviews as good ones and do this quickly. It is also important to report false reviews to make your listing as clean as possible.

Important: You cannot report bad reviews only because they make your business look bad. There must be a clear violation of policy for a review to be removed.

Biggest Factor: It's One You Cannot Control

Now you have completed the arduous work of optimizing your listings. You have great reviews, all the information is correct, so now when anybody searches for your service/product locally (i.e., "near me"), you're guaranteed to be listed. Unfortunately, this is not the case because the single most important factor is proximity – How close is your business to the searcher.

There is so much more to voice search that we did not cover here. We only touched on local searches, which tend to be map searches. Voice search is rapidly evolving, so you will want to revisit this area of your campaigns frequently.

Is Your Website Ugly?

As you may have guessed, there is a story behind this chapter. Every day at 8 AM, we meet up and talk about what is going on. We share challenges, discoveries and talk about what's next. In one meeting, we discussed the need to place a "Your baby is ugly" call. An ugly baby call is when we must tell a client the website they are heavily and personally invested in is data ugly.

Jackie, our Marketing Strategist, was discussing a new account that had just finished its first marketing cycle, and the results were ugly. Google Ads received the targeted traffic level, and the search query results were clean, but the results were ugly. The conversion rate was a terrible 0.16%, with a goal of 2%.

We were concerned when we first saw the site, but the client told us it worked for them, and we always trust what clients say until data proves otherwise. We held our input and waited for data. Our first impressions have been wrong before, and we always trust our clients. We have seen many sites that we thought were ugly that produced superior results for clients, making us smarter.

With data in hand, we needed to have the dreaded "Your Baby (aka Website) Is Ugly" client call. While there was nothing funny about an ugly website, we did have a good laugh with the **"Your Baby is Ugly"** twist. As a Content Scientist, I set out to find a website ugliness formula, and this chapter is about that journey.

The Ugly Formula

Our Designer, from an artist's perspective, said:

"Beauty is in the eye of the beholder, and there is no formula for beauty."

As the Content Scientist, I called BS on this. While the emotion of art is difficult to put a number to, a website's performance is not. As a business, the beauty standard is profitability. Not only can you put a number to it, but it is a **best practice** to do so. Beauty comes in two forms:

- ➤ CPA (Cost Per Action) – expressed in dollars
- ➤ VOA (Volume of Actions) – expressed in units
- ➤ Beauty is positive – the bigger the profit, the prettier the site
- ➤ Ugly is negative – the bigger the loss, the uglier the site
- ➤ Here is the math:
  - ○ ugly$ = Value – Cost – CPA
  - ○ ugly Units = Actual – Target
  - ○ **It is this simple!**

The action of the visitor creates value. Value can take many forms, but ultimately it must be expressed as a dollar value in the formula. For Executives that have never been through this before, the conversation of getting to this value can be difficult. The difficulty comes from them trying to get to a specific exact value. Marketing data is generally right but never perfectly accurate. Marketing data is made up of averages and estimations based on imperfect tracking. To help lead clients through this, we ask them for the average order size and the average cost of goods. Use the P&L to get the total sales number from last year and divide it by the number of orders last year. That will give you a number that is accurate enough. Sometimes there are non-repeating orders that must be removed from the data, but that is rare. If the business is new, then they must guess.

Cost is all direct costs associated with the value and can be gross or net. Gross is the direct cost of the product, and net considers the contribution to overhead, but this can get wildly complicated. It is accurate enough to take the total marketing expense from last year and divide it by the number of orders. Then add the direct product costs, and you have the net cost.

CPA is the Cost Per Action, and it is what the business sets as the value of the action taken. Different businesses talk about this using a variety of terms like Customer Acquisition Cost or Cost Per:

➢ Orders
➢ Leads
➢ Subscriptions
➢ Engaged Visitors
➢ Returning Visitors
➢ Minimum Time On-Site

Proper tracking will give you the base CPA, but you need to modify that by the performance of the rest of the sales pipeline. Look at the performance of any remaining steps from the CPA to the final business. In many cases, this includes the performance of your Customer Service and field technician. As an example:

Description	Amount
Phone Call CPA	$65
Customer Service Conversion to Appointment	75%
Technician Close Rate	85%
Average Ticket	$450

Average Ticket Gross Profit	50%

Some simple math tells us that our final CPA is $101.96 to produce $225 in profit. Here is how we got there. $65 / .75 / .85 = $101.96 resulting in a gross net profit $123.04 or 27%.

CPA above the CPA Target is Ugly

CPA below the CPA Target is Beautiful

The line between Ugly and Beautiful is your CPA Target.

Set a Reasonable CPA

We have client conversations all the time to help them set CPA Targets. We ask the question, **"How much would you pay for a lead?"** and often, the first answer we get is **"As little as possible."** Certainly, a reasonable answer in their mind, but not in their market. Every business wants expenses to be as low as possible, and initially, everyone thinks of CPA as an expense. CPA is an investment, not an expense. The lower the CPA, the lower your volume will be. Cost and Volume are strongly connected with a string-like relationship. It is impossible to pull on one end of a string without moving the other end. Take the Target CPA multiplied by the Target VOA, and that is your Budget.

$$Budget = CPA * VOA$$

Call to Action

Acting on this data often requires forgiveness and separation from the past. Many clients are hugely vested in Ugly websites because they were once beautiful and dominated their markets. They are immensely proud of their ugly baby and defend it fiercely. There are a few sites that age like a fine wine and should not be replaced. Make sure you know which one you have and let the data tell its story. Over their lifetime, websites just need maintenance and tune-ups. However, at some point, it will need to be scrapped so the next generation can come of age.

Sales are 95% Emotional

Sales are marketing are driven by emotions, not logic. Let me say that again **"Driven by emotions, not logic."** It is not the low price or the best product that wins the day. It is how we connect to the consumer at an emotional level. People propose that the price, product features, or value, is how you win a market. We propose to you that it is unsupported by the facts of the market. You can find outlier data that proves the logical argument, but emotion, not logic, makes the sales and marketing world go round and round.

The emotions that drive most of the market are **greed and fear.** 95% of Sales are emotional, not logical. People make up their decision and then use logic to justify their actions. Focus on your what your customer wants and use what they need to help them justify their decision. Sales can be grouped into two major types:

1. Transactional
2. Relational

Transactional sales are all about closing the sale, and the world is full of examples of these. In the opinions of the authors, far too many businesses think like this, and it leads to bad decisions. In transactional sales, profitability is measured on only the current transaction without considering building the brand. Products are sold on landing pages with a "My way" or the "highway" design. This type of landing page and expects to get a sale from that. The consumer either buys or does not buy the product with no fallback or secondary offer. In this model, the consumer will NOT return later to buy. While not 100%, most transactional sales are products, not services.

Relational sales involve traffic that returns multiple times before the first transaction happens. The profitability of relational sales must consider the life span of the consumer, not the transaction. Relational sales involve recurring revenue such as subscription services or SaaS (Software as a Service). If you count only the first sale, most subscriptions and SaaS sales are not profitable, but we can assure you that they are. After the first few months, the sales costs have been recovered, and the profits increase.

The state of mind of the customer is the next dimension to understand. We have two states to consider:

1. Emergency
2. Non-Emergency

Emergency traffic is what it sounds like. The consumer is under pressure to procure a product or service to solve an immediate problem. The classic example is when you call a Plumber because stuff is spewing on the floor. Emergency traffic exists on a spectrum of pressures from minor to life-threatening.

Non-Emergency traffic is a planned purchase with much less pressure. These sales have a quote, bid, sales cycle of a variable length, and that length is much longer than most businesses think. Many customers have told us their sales cycle is two weeks, only to discover that reality was closer to a year. Customers do not intend to lie to us, but they base the two weeks on the orders they can see. Many orders can have 20 or more marketing touches or steps over a long time. These are incredibly difficult to track and measure, so they can only observe and count the last few steps.

Think of non-emergency sales like a basketball game. From one end of the court to the goal, there are dozens of dribbles and passes with one final shot, and there would be no final shot without the dribbles and passes. In marketing, dribbles and passes are called "Touches." Without the twenty touches, there would be no sale. That is the part of the marketing game that many business people fail to observe. They want the sale to go through a structured, logical flow, but sales are neither structured nor logical. Like basketball, marketing requires a team with constant adjustments as the sale works its way to the close.

Things that cause us to Act

There is no doubt that greed and fear are the big ones, but there are lots of other things that can move us to action, and the section in the first part of the book labeled "Call to Action" discusses more of these.

About the Authors

This book was written by the Team at Systems & Marketing Solutions. We have been involved with Internet marketing since 1994 and Google  Ads since 2003. We have helped hundreds of SMBs manage their Google Ads account to produce cost-effective results.

Bob Dumouchel

 Bob is an Internet Pioneer with his first e-commerce sale in October 1994. He has thrived through Y2K, The Internet Bubble, Sept 11th, 2008-9 Recession, and now COVID-19. As a consultant, Bob has helped hundreds of businesses in dozens of industries to understand how to leverage technology in Sales & Marketing and create website experiences worth talking about.

Bob started his career as a computer programmer. Since 1988, Bob has served as the CEO of eight technology startups, leading them from concept to operational profitability. Selected twice to the Inc 500 List (92 #247, 93 #467), his companies have been Premier Business Partners with Google and IBM. Bob started working with Google Ads in 2003 and actively consults on accounts with difficult PPC challenges. Bob and the whole staff at SMS were in the first group of 50 professionals certified in AdWords (today called Google Ads).

After turning the SMS CEO role over to Josh Erdman in 2017, Bob serves as a Content Scientist. He spends his days conducting data experiments, creating marketing innovations, and writing about discoveries. Bob has published hundreds of sales and marketing articles and three books. He serves as an advisor to the most complicated and demanding client accounts within SMS.

Priyanka Singh

Priyanka is a Digital Sales Engineer with a background in statistical analysis. She graduated from Cal Poly earning a BS in Statistics and is the contributing author for the 2021 version of the book "The Mission is Sales." She works closely with Digital Sales Engineers and Strategists. Priyanka earned her Google Certifications for Search and Display and is currently working on advanced industry certifications. Priyanka also served as a primary book editor.

Josh Erdman

Josh is the current CEO of SMS and oversaw this book project and the staff support for this initiative. Josh has over 20 years of experience in the technology industry. Josh is the driving force behind our innovations in Digital Marketing. Josh is a passionate speaker, drawn to learning, and excited to inspire.

Got A Question? Call Us

We are always open to a conversation about digital marketing. If you have questions or suggestions for new areas, you would like added to the next book, grab a phone and give us a call.

(800) 272-0887

Or visit our web site at www.smsrd.com

Index

Glossary

The meaning of some words can vary from expert to expert, and this glossary is provided to help you understand the words within the context of this book. It is a big, complicated world out there, and you will find variations in the exact meanings.

Ad Spend: The amount spent on ads.

Analytics: Analytics consists of a program that records and helps you interpret the visitor activity on your site. When we refer to analytics, we are specifically talking about Google Analytics, a free analytics program.

Average Position: Average position is a metric that indicates what position your ads are typically in. Because of the fluid nature of PPC, your ads won't be in the same position every time, so you need to rely on the average position to get the most accurate measurement of your position on a given search.

Average Position Overlap Rate: How often you and the other advertiser show up in the same results.

Bounce Rate: Bounce rate is a measurement within Google Analytics that tracks how many visitors leave your site immediately without viewing any other pages. This metric is helpful when determining the value of traffic from certain keywords and sources. It can also be an early warning to deeper problems in the site due to content, design, or poorly targeted marketing.

Campaigns: Campaigns are the highest level of summarization within the Google Ads Account. Campaigns have broad controls of ad delivery. Common items managed at this level are the budget, networks selection for ad delivery, delivery schedule, delivery methods, types of bidding, position preference, ad serving, and regional delivery areas. Campaigns are the highest level of negative keyword management and provide access to category exclusions.

Clicks: Clicks happen when the searcher interacts with your advertisement.

Click Fraud: Click fraud is an instance where a competitor or someone with a financial interest in clicking your ads does so to either deplete your budget or make money for themselves via Adsense. Google Ads has numerous monitoring filters to detect click fraud and refund clicks that it feels are not legitimate. While this is effective, you should watch your account for unusual activity and report problems as you see them.

Display Network: The Content Network is an ad service within Google Ads. The Content Network places your advertisements on various websites worldwide based on the content on that site. In the past, this was called the Content Network.

Conversions: Conversions are when the searcher performs a process that you have deemed to have business value. Conversions take many different forms, including online sales, email list subscription, or even a phone call. Conversion tracking is a challenge, and it is not just a technical problem because conversions can happen online or offline.

CPA: (Cost Per Acquisition - AKA Cost per Conversion) Cost per lead from PPC activity

CPC: (Cost per Click) In PPC, you pay each time your ad is clicked.

CTR: (Click-through Rate) The percentage of people who click on your ad after seeing it.

Direct Traffic: People who type in your website address directly into their browser

GDN: Google Display Network - see Display Network above.

Image/banner Ad: An Image Ad is a graphic advertisement served via Google Ads to the Display Network.

Impressions: An impression is a display of the ad, and not all impressions are created equal. It is the quality and quantity of impressions that start the flow through the system.

Impression Share: The percentage of time that the advertisement shows up when eligible to show.

Landing Page: A landing page is where your advertisement directs the user when they click on the ad.

Marketing Collateral: A collection of media used to support the sales of a product or service. Historically, the term "Collateral" specifically referred to brochures or sell sheets developed as sales support tools. In digital, it is often called assets and includes pictures, logos, and other items that are commonly reused in your marketing.

Organic Traffic: Organic Traffic is search traffic that is not paid for. This traffic is typically the clicks in the section under the sponsored ads.

Page Score: Page scoring is the total of the page attributes. When all the scores are sorted in ascending order, the order of the pages is the rank.

Quality Score: Quality Score is a number calculated by Google that determines the relevancy of your keyword. It considers your website, your ad copy, your CTR, and other factors.

Referral Traffic: People who come to your website via a link on another website.

Sales Funnel: The buying process that companies lead customers through. They are typically divided into several steps, which differ depending on the sales model.

Search Engine Optimization (SEO): The process of improving the position on the SERP by items other than paid advertising.

Search Engine Results Page (SERP): The search engine's results page is created by a search engine based on the search query entered by the searcher.

SMB Small and Medium-sized businesses. Generally, this is less than 500 employees.

Targeting: In digital advertising, advertisers can target certain demographics (geography, gender, interests, job position, industry, and others). The demographics you can choose from vary between which ad network your ads are active in. Facebook, for example, allows targeting by age, gender, geography, and even interests.

Tactic: A specific targeting method used within a campaign to gain the attention of your prospects or customers.

Behavioral Targeting: Targeting based on prior behavior of the user. Behavioral targeting is particularly successful in Social or Display Networks.

Contextual Targeting - Display Ads shown in the context of other content. For example, a hotel may show a display ad next to an article on travel.

Demographic Targeting - Use demographic targeting to show display ads specifically to those people who are most likely interested in your business within your service area. Demographic targeting is particularly successful when used to refine another targeting type.

Geographic Targeting - Geographic and language targeting lets you display your ads by language, region, even postal code.

Interest Categories - Adjust keywords bids for audiences whose online behaviors show they share a common interest with your industry.

Site Remarketing - Bring your audience back! People who have already visited your site are shown your ads as they browse other sites on the Google Display Network.

Volume of Actions (VOA): This is the volume of actions related to the CPA. The VOA and CPA are closely related, with the difference being the cost per click (CPC).

www.ingramcontent.com/pod-product-compliance
Lightning Source LLC
Chambersburg PA
CBHW071238050326
40690CB00011B/2175